Bioenvironmental Systems

VOLUME I

VOLUME II

VOLUME III

VOLUME IV

Bioenvironmental Systems

Volume IV

Editor

Donald L. Wise, Ph.D., P.E.

President
Cambridge Scientific, Inc.
Belmont, Massachusetts

CRC Press, Inc.
Boca Raton, Florida

Library of Congress Cataloging-in-Publication Data

Bioenvironmental systems.

Bibliography: p.
Includes index.
1. Sewage sludge digestion.   2. Organic wastes.
3. Biomass energy.   4. Biotechnology.   5. Methane.
I. Wise, Donald L. (Donald Lee), 1937-
TD769.B56   1987   628.3′51   86-29936
ISBN 0-8493-4512-X (set)

This book represents information obtained from authentic and highly regarded sources. Reprinted material is quoted with permission, and sources are indicated. A wide variety of references are listed. Every reasonable effort has been made to give reliable data and information, but the author and the publisher cannot assume responsibility for the validity of all materials or for the consequences of their use.

Direct all inquiries to CRC Press, Inc., 2000 Corporate Blvd., N.W., Boca Raton, Florida, 33431.

International Standard Book Number 0-8493-4512-X (set)
International Standard Book Number 0-8493-4513-8 (v. 1)
International Standard Book Number 0-8493-4514-6 (v. 2)
International Standard Book Number 0-8493-4515-4 (v. 3)
International Standard Book Number 0-8493-4516-2 (v. 4)

Library of Congress Card Number 86-29936
Printed in the United States

PREFACE

Bioenvironmental Systems is an especially valuable reference text consisting of contributed chapters in which the most promising and active bioenvironmental research and development programs around the world are described. The authors of these contributed chapters are very serious and dedicated technologists who are exploring practical solutions to environmental problems. This important text has as its major theme the bioconversion of organic residues. This major theme primarily encompasses the field of anaerobic methane fermentation. The text is intended to present a comprehensive overview of the most practical research programs that are being carried out in this emerging field of international significance. Due to the fact that both research investigations and practical systems development technology, with emphasis on methane fermentation, have been under development at key sites around the world, great care has been taken to include chapters from an international perspective. Furthermore, as a perusal of the chapter titles will indicate, an emphasis has been made to address both the important research and practical aspects of the work on bioenvironmental systems. It is to be noted that each chapter included in this text is the work of a particular individual or group. There are no multiple chapters by more than one author or group. Thus, each of the included chapters most often reflects the dedicated career efforts of these workers. Each contributed chapter is presented on a stand-alone basis so that the reader will find it helpful to consider only the theme of each chapter. However, there is the unifying theme will all chapters of addressing bioenvironmental systems research and development. A reader of this text just entering the field will find it provides an excellent state-of-the-art presentation of the international import of work on bioenvironmental systems, with emphasis on methane fermentation. A reader of this text who has experience in this field will find it to be essential for assessment and referral of this increasingly valuable area of technology.

THE EDITOR

Donald L. Wise, Ph.D., P.E., is Founder and President of Cambridge Scientific, Inc., Belmont, Massachusetts. Dr. Wise received his B.S. (magna cum laude), M.S., and Ph.D. degrees in chemical engineering at the University of Pittsburgh. Dr. Wise is a specialist in process and biochemical engineering as well as advanced biomaterials development. During his career he has managed a series of programs to develop processes for production of fuel gas, liquid fuels, and organic chemicals from municipal solid waste, an array of agricultural residues, and a wide variety of crop-grown biomass, especially aquatic biomass. Dr. Wise has also been primarily responsible for the initiation of development work on fossil fuels such as peat and lignite to gaseous fuel, liquid fuels, and organic chemicals, and he also originated work on the bioconversion of coal gasifier product gases to these products. Dr. Wise initiated a program to establish the engineering feasibility of converting large-scale combined agricultural residues to fuel gas by the action of microorganisms, a project ultimately involving joint effort with research workers in fifteen countries around the world.

CONTRIBUTORS

Emer Colleran, Ph.D.
Statuatory Lecturer in Microbiology
Department of Microbiology
University College
Galway, Republic of Ireland

Vidar Friis Larsen, Ph.D.
Lecturer in Process Biotechnology
Department of Chemical and Process
  Engineering
University of Strathclyde
Glasgow, Scotland

Ian S. Maddox, Ph.D.
Senior Lecturer
Department of Biotechnology
Massey University
Palmerston North, New Zealand

Declan Maher, B.Sc.
Research Associate
Department of Microbiology
University College
Galway, Republic of Ireland

Dr. Michel Paquot
Department of Technology
Faculte des Sciences Agronomiques
Gembloux, Belgium

Laura Postils
Servei d'Agricultura i Ramaderia
Diputacio de Barcelona
Barcelona, Spain

Morton W. Reed, Ph.D.
Associate Professor
Department of Textile Engineering
Auburn University
Auburn, Alabama

Joan Rieradevall
Chemist
Laboratori Agrari
Diputacio de Barcelona
Barcelon, Spain

Antonio Ruera
Agronomist Engineer
Servei d'Agricultura i Ramaderia
Diputacio de Barcelona
Barcelona, Spain

Wen K. Shieh, Ph.D.
Assistant Professor
Department of Civil Engineering
University of Pennsylvania
Philadelphia, Pennsylvania

Philippe Thonart
Charge de Cours
Department of Technology
Faculte des Sciences Agronomiques
Gembloux, Belgium

Montsenat Vicente
Servei d'Agricultura Ramaderia
Diputacio de Barcelona
Barcelona, Spain

TABLE OF CONTENTS

Chapter 1

THE ENVIRONMENTAL SIGNIFICANCE OF THERMOSENSITIVE PLASMIDS OF THE H INCOMPATIBILITY GROUP

Declan Maher and Emer Colleran

TABLE OF CONTENTS

I. INTRODUCTION

The discovery and development of antibiotics and of safe and reliable synthetic antimicrobial agents in the first half of this century has revolutionized both human medicine and animal husbandry. Unfortunately, world-wide usage of chemotherapeutic drugs for the control of infections caused by microorganisms has been accompanied by the emergence and widespread dissemination of drug resistance among both pathogenic and nonpathogenic bacterial species. Currently, microbial drug resistance has reached such epidemic proportions that it constitutes a significant threat to the continued successful treatment of infectious disease.

Drug resistance, particularly among Gram-negative bacteria, is usually associated with the presence of extrachromosomal hereditary determinants known as R-factors. Although it is generally agreed that R-factors existed before the development of modern antibacterial drugs, the widespread use and abuse of these drugs has resulted in a vast increase in the incidence of drug resistance caused by R-factors. R-factors were originally described by Watanabe[1] and may be defined as autonomously replicating DNA entities (i.e., plasmids) which carry the genetic determinant(s) for resistance to one or more antimicrobial drugs. R-factors may be transferred either by conjugation or by transduction, although the significance, in nature, of phage-mediated transfer is unclear.[2] R-factors are promiscuously transferred by conjugation both among the *Enterobacteria* and to other Gram-negative bacteria outside of the Enterobacteriaceae.[2] The frequency of transfer of R-factors from natural isolates to suitable recipient bacteria is generally low — 10^{-4} per R^+ cell in 1 hr of mixed incubation[2] — and is due to the normally repressed state of the R-factor sex pili. Studies with a derepressed F-like R-factor (RI) showed that transfer at 37°C proceeded within 5 min of mixing of the two populations and was complete within 15 min.[2] In addition, the recipient cell could replicate the incoming R-factor DNA, synthesize the appropriate R-pilus and other necessary proteins, and function as a donor in another round of conjugal transfer within 15 min of receipt of the R-factor. It is clear, therefore, that transferable drug resistance can spread like wildfire under ideal in vitro conditions throughout a susceptible bacterial population in a relatively short time.[3] Without doubt, transfer of drug resistance also occurs in vivo, both in animals and man,[4] but the frequency with which it occurs and its contribution towards the general prevalence of antibiotic resistance are matters of considerable dispute. Lacey[5] concluded that transfer of drug resistance rarely occurs in the normal gut but may be detected more regularly under certain conditions, such as would prevail, for instance, during antibiotic therapy. Even if such occasions are rare, transfer in vivo may have far-reaching effects, especially if drug resistance is transferred to a pathogen and opportunity for its spread to other hosts subsequently arises.[4] Gardner and Smith[6] suggested that environmental selection by antibiotics and colonization of patients with resistant organisms have far greater epidemiological significance than R-factor transfer in the GI tract. If this is true, the extent of antimicrobial drug usage for disease control in human medicine and for therapeutic, prophylactic, and growth promotion purposes in agriculture is of major importance, as are the routes for cross-infection in the human population or for cycling of resistant strains between animals and man.

The majority of studies on R-factor prevalence, ecology, and transferability have been heavily biased at 35 to 37°C because of the public health implications of these transmissible elements in human and veterinary medicine. R-factor transfer in voided feces, in sewage or other wastewater treatment plants, in fecally contaminated water bodies, or in sediments has not been generally regarded as a significant contributory factor to the spread of drug resistance because of the constraints put upon such transfer by environmental temperature optima.[7] The discovery, however, of R-factors which are temperature sensitive for transfer, i.e., transfer optimally at temperatures below 35°C, significantly increases the potential epidemiological importance of environmental R-factor transfer and raises questions as to the need for a re-evaluation of standards for wastewater treatment systems and effluent discharge into natural waters.

II. CHLORAMPHENICOL RESISTANCE IN *SALMONELLA* AND THE DISCOVERY OF Inc H PLASMIDS

The usage of chloramphenicol as the drug of choice for the treatment of typhoid fever was not compromised, prior to 1972, despite the presence of plasmid-mediated chloramphenicol resistance in the enterobacteria and its demonstrated transferability at high rates in vitro to *Salmonella typhi*. However, in 1972 to 1974 a major epidemic of typhoid fever in Mexico (>10,000 cases) was shown to be caused by a chloramphenicol-resistant strain of *S. typhi* which carried a plasmid mediating resistance to streptomycin (S), sulfonamides (Su), and tetracycline (T) as well as to chloramphenicol (C).[8] The plasmid was shown to belong to a new incompatibility group, designated Inc II by Grindley and co-workers,[9] and to be thermosensitive for transfer, i.e., it transferred into *Escherichia coli* K12 in overnight crosses at 30°C with a frequency approaching unity whereas transfer was as low as 10^{-6} in similar overnight crosses at 37°C.[8] Further work resulted in the subdivision of group H plasmids into H_1 and H_2. The Mexican CSSuT plasmid is regarded as the prototype of Inc H_1. Inc H_2 are incompatible with Inc H_1 and their DNA shows little homology with that of H_1, although they display considerable similarity in molecular size, $G + C$ content, and transferability.[8] The prototype H_2 plasmid was identified from a strain of *S. typhi* originating in Spain.[10]

In the next few years, *S. typhi* isolates carrying H_1 plasmids with CSSuT determinants were found to be the causal agents of separate typhoid epidemics in India, South Vietnam, and Thailand (Table 1). Although the Mexican and Indian outbreaks were clonal epidemics associated with a single Vi-phage type, nine different Vi types were involved in the South Vietnam and Thailand epidemics.[8] Since it is inconceivable that different Vi types were

Table 1
GROUP H₁ R-FACTORS IN *S. TYPHI* PRIOR TO 1975[8]

Country of origin	Host	Year of isolation	Drug resistances	No. of isolates examined
Mexico	Human	1972	CSSuT	8
India	Human	1972	CSSuT	26
South Vietnam	Human	1972—1974	CSSuT	70
			CSSu	1
Thailand	Human	1973—1974	ACSSuT	5
			CSSuT	13

Note: Drug resistance abbreviations are A (ampicillin), C (chloramphenicol), S (streptomycin), Su (sulfonamides), T (tetracycline).

common enough to reach the cellular numbers required for direct R-factor transfer to occur
between them, it must be assumed that each strain of the Vi types represented had acquired
its H₁ R-factor in a discrete event, probably from nonpathogenic intestinal organisms such
as *E. coli*.[8]

The association of chloramphenicol resistance with Inc H plasmids was also demonstrated
in *S. typhimurium* and in other *Salmonella* serotypes. H₁ R-factors coding for AT, AK, or
AKT resistances (A = ampicillin; K = kanamycin) were isolated by Anderson and co-
workers[8] from *S. typhimurium* phage type 193 during outbreaks of animal and human
infection in Singapore and Malaysia. In Britain, the detailed studies of Threlfall and co-
workers[11-13] on *S. typhimurium* infection of calves from 1974 to 1980 clearly illustrated the
tendency of specific *Salmonella* strains carrying H plasmids to be distributed in a clonal and
epidemic fashion. Prior to 1977, several substantial milk-borne outbreaks of *S. typhimurium*
were associated with phage type 204 which was resistant to sulfonamides and tetracycline.
In June 1977, a strain of type 204 caused epizootic infection of calves in a farm in Leices-
tershire. Genetic studies demonstrated that the new type 204 strain had acquired a H₂ plasmid
mediating CSSuT resistances.[13] Subsequently, in March 1979, a strain of *S. typhimurium*,
designated 204c, was identified in calves. The strain was resistant to CSSuTTm (Tm =
trimethoprim) and genetic studies showed that the H₂ plasmid had acquired a trimethoprim-
resistant transposon. Type 204c quickly became established in calves in Britain and by
October 1979 had acquired further resistance plasmids (non-H types) specifying ampicillin
(A) and kanamycin (K) resistance. By December 1979 the predominant R-type in 204c was
ACKSSuTTm.[14]

Type 204a, R-type SuT, appeared in October 1978 and subsequently spread in cattle in
Britain, ultimately entering the human food chain.[14] By 1980 it had acquired a H₁ plasmid
mediating resistance to CKSSuT. These elegant epidemiological studies clearly showed,
therefore, that plasmids of the Inc H group were selected by the prophylactic and therapeutic
use of drugs in calf husbandry in intensive units and spread extensively throughout the U.K.
by the large-scale movement of calves through dealers' premises and markets.[15] Apart from
the economic implications to the cattle industry, the widespread dissemination of multiresistant H plasmid-carrying *S. typhimurium* in bovine sources had considerable clinical significance. Threlfall reported 505 human infections with Inc H-containing *S. typhimurium*
strains (289) and non-Inc H strains (216) between 1977 and 1980 in Britain.[13] Although the
majority of patients exhibited symptoms of mild to moderate enteritis, extraintestinal spread
resulting in septicaemia or meningitis occurred in ten cases. Obviously, the multiresistant
nature of the causative *S. typhimurium* strains severely limited the clinicians' choice of drugs
for therapy.

These and other studies clearly illustrated the medical and veterinary significance of
plasmids of the Inc H group. The origin of these plasmids, the significance of their tem-

perature sensitivity for transfer, their prevalence in other Gram-negative bacteria and in man and animals from different geographical areas, the molecular biology of their transfer system and the other genomic determinants encoded by their high molecular weight DNA were, however, largely unknown and have attracted considerable research during the past 10 years.

III. CLASSIFICATION OF H PLASMIDS

A. Incompatibility Typing

Plasmids may be grouped according to their ability to co-exist stably in the cytoplasm of their bacterial host.[16,17] At least 26 different incompatibility (Inc) groups are known among the plasmids of enteric bacteria. Generally, plasmids within a group are excluded symmetrically, only one of any given pair being stably maintained.

As described above, early studies of Inc H plasmids designed to ascertain the degree of DNA homology within the group[9,18] and the ability, or lack of it, of members within the group to displace the F-factor from a recipient cell[19] led to the subdivision into Inc H_1 and H_2 types. Taylor and Grant[20] later renamed these subgroups as Inc H1 and Inc H2. A third subgroup, designated Inc H3, was subsequently recognized. The studies of Bradley and co-workers[21] revealed further anomalies and led to a reclassification of the entire group as detailed below.

Two separate H incompatibility groups are now recognized and designated Inc HI and Inc HII.[21] The former are all thermosensitive for conjugal transfer and are large molecules with molecular weights of 100 Mdaltons or more. Inc HI plasmids are subdivided as follows.

1. Inc HI1

These plasmids were formerly referred to as Inc H_1[8] or H1.[20] Plasmids in this subgroup display strong incompatibility with each other and a weaker incompatibility with members of the Inc HI2 and Inc HI3 subgroups.[20] They are generally compatible with Inc HII plasmids[21] and display asymmetric incompatibility with F, the F-factor always being displaced.[19]

2. Inc HI2

These plasmids were formerly referred to as Inc $H1_2$[8] or Inc H2.[20] Plasmids in this subgroup display strong incompatibility with each other and a weaker incompatibility with members of the Inc HI1 and Inc HI3 subgroups.[20] They are generally compatible with Inc HII plasmids[21] and with the F-factor.[19] Plasmids, which were isolated from *Serratia marcescens* and which were originally designated as Inc S,[22] have been shown to belong to the HI2 subgroup.[20]

3. Inc HI3

Formerly known as Inc H3, this subgroup presently contains only one known plasmid, MIP 233,[23] which is incompatible with plasmids of both the HI1 and HI2 subgroups. Differentiation from the other HI subgroups is based on DNA-DNA membrane filter hybridization studies.[24]

The pili specified by Inc HI1 and Inc HI2 plasmids are serologically related H pili; those specified by Inc HI3 have recently been shown to be serologically different from all known pilus types, including H pili.[175]

4. Inc HII

Two unique plasmids from *Klebsiella aerogenes* were found by Bradley and co-workers[21] to be incompatible with one another, to surface exclude one another, to determine the constitutive production of H pili, to have molecular weights of the order of 100 Mdaltons, and to code for resistance to tellurium. These plasmids do not show surface exclusion of, or by, plasmids of the Inc HI group and have been placed in a separate Inc H group, HII.

Plasmids of this group are not temperature sensitive in their transfer function and transfer at similar high frequencies at 30 and 37°C.[21] A third plasmid, pHH1532b-1, isolated in 1978 from *K. aerogenes,* was subsequently reclassified and included in this group.[21]

The nonthermosensitive plasmid R831b, which was originally described as belonging to Inc H,[25] has been reclassified as Inc M.[26]

The classification of plasmids suspected of belonging to Inc H is often laborious and time-consuming because of shared antibiotic resistance markers with standard Inc H plasmids. The carriage of less common phenotypic markers such as sucrose- or lactose-metabolizing genes by some H plasmids has increased the ease and rapidity of typing unknown R-factors.[23] Plasmid classification by integrative incompatibility[27] has also been effectively used since it directly differentiates between Inc HI1 and Inc HI2 plasmids when mating is carried out on membrane filters.[28] Size differences between Inc H plasmids are rarely sufficient to permit determination of incompatibility by visualization of band loss in gels.[29]

B. Fertility Inhibition (fi)

The classification system used initially for R-factors was based on inhibition of the fertility function of the F-factor.[30] R-factors which inhibited the function of F in F^+, Hfr, or F' cells were designated fi^+ and those which lacked this ability were designated fi^-.[31] It was subsequently discovered that bacteria harboring fi^+ R-factors were lysed by F-specific phage while cells containing fi^- R-factors were lysed by I-specific phage.[32] Subsequent studies, however, showed that the latter phenomenon was only true for I-like R-factors and that a variety of fi^- R-factors existed which did not allow lysis of their hosts by I-specific phage, suggesting that these R-factors encoded a transfer system specifying a different pilus type. The incompatibility system described above was subsequently adopted and this permitted a more valid classification of R-factors into different Inc groups.

Because of the high degree of homology between plasmids within an Inc group, the fi character might be expected to be conserved. However, several instances of fi^+ R-factors within a predominantly fi^- Inc group have been reported and these plasmids probably inhibit F-pilus formation by mechanisms different from that of F-like R-factors.[33] Plasmids belonging to the Inc H group are typically fi^-. Although the standard HI1 plasmid, R27, was initially designated as fi^+, this was later shown to be attributable to elimination of the F-factor from the F^+ recipient.[34] Several other fi^+ Inc H plasmids have been reported by other workers.[35-37] In these cases, inhibition of the fertility function of the F-factor was expressed in Hfr cells where the F-factor is integrated in the chromosome.

IV. THE PREVALENCE OF Inc H PLASMIDS IN THE ENTERIC BACTERIA AND THEIR GEOGRAPHICAL DISTRIBUTION

Since literature reports on the prevalence of Inc H plasmids are largely limited to studies of *Salmonella* serovars in Europe, Japan, and North America, it is difficult to present a balanced appraisal of the species distribution and world-wide prevalence of Inc H. The situation is further complicated by the fact that dominance of a particular plasmid in a given locality at any time may simply reflect local usage of antibiotics at that time.[38]

A. Geographical Distribution

Anderson reported in 1977 on the world-wide distribution of plasmid types in salmonellae, particularly in *S. typhi* and *S. typhimurium.*[39] Considerable regional difference was noted in *S. typhimurium,* with plasmids of different Inc groups showing definite regional predominance (Table 2). In Northwestern Europe, Inc I_1 plasmids predominated in *S. typhimurium* of human and animal origin and plasmids of the Inc H type were not detected in the 1748 isolates examined. In Southern Europe, group HI2 plasmids were shown to predominate in

Table 2
COMPATIBILITY GROUP DISTRIBUTION OF RESISTANCE PLASMIDS IN *S. TYPHIMURIUM*, 1968—1976[39]

Area of origin (no. of cultures examined)	Source of S. typhimurium	Dominant Inc groups (% of transferable plasmids)							
		I_1	4_1	N	F_{II}	4_2	HI1	HI2	$F_I me$
Northwest Europe									
Belgium (1203)	Belgium: 6.2% human								
Belgium (1203)	Belgium: 6.2% human								
France (73)	93.8% animal	73.6	14	7.3	—	—	—	—	—
Germany (460)	Others: human								
Finland (12)									
Southern Europe									
Greece (98)	Italy: animal								
Spain (60)		29.6	—	9	15.9	—	—	26.5	—
Italy (3)	Others: human								
Portugal (71)									
South America									
Argentina (93)									
Peru (5)									
Brazil (372)									
Uruguay (73)	Human	—	—	84	1.9	—	—	—	—
Chile (135)									
Venezuela (12)									
Paraguay (6)									
North America									
Canada (336)	Canada: 50.3% human	44.5	—	8.3	11.6	—	—	—	—
Mexico (100)	49.7% animal / Mexico: human							30.6	—
Australia (193)	Human: 57%	41.0	—	4.8	5.7	39	—	2.9	—
New Zealand (165)	Animal: 43%								

Table 2 (continued)
COMPATIBILITY GROUP DISTRIBUTION OF RESISTANCE PLASMIDS IN *S. TYPHIMURIUM*, 1968—1976[39]

Area of origin (no. of cultures examined)	Source of S. typhimurium	Dominant Inc groups (% of transferable plasmids)							
		I_1	4_1	N	F_{II}	4_2	HI1	HI2	F_1me
Southeast Asia									
India (3)	Singapore: 46.2% human 53.8% animal								
Singapore (93)									
Malaysia (77)	Malaysia: 97.4% human 2.6% animal	16.6	—	—	12.2	—	67.2	—	1.1
Philippines (8)	Others: human								
Africa									
Ghana (33)	Ghana: 18.2% human 81.8% animal								
Liberia (5)		44.8	—	—	—	—	—	20.7	34.5
Kenya (20)									
Tanzania (1)	Others: human								
Middle East									
Iran (142)									
Iraq (2)									
Israel (163)									
Jordan (1)	Human	2.0	—	—	—	—	—	23.7	72.0
Kuwait (2)									
Syria (1)									
Turkey (2)									
Other (2)									

S. typhimurium isolates from human sources in Spain and Portugal whereas Inc I_1 was the most common plasmid type in human isolates from Greece. Anderson concluded that group HI2 plasmids were endemic in *S. typhimurium* in Spain. No Inc H plasmids were obtained in isolates from severe pediatric epidemics in South America (Table 2) where N plasmids dominated. In North America, animal and human isolates from Canada carried mainly HI2 plasmids. Despite the association of Inc HI1 with the Mexican typhoid epidemic in 1972 to 1974, plasmids in *S. typhimurium* from human sources in Mexico were predominantly Inc I_1. No Inc H plasmids were identified in isolates from Australia and New Zealand whereas Inc HI1 was the dominant plasmid type in isolates of human and animal origin in Southeast Asia. In Africa, Inc HI2 contributed 21% of the transferable plasmids studied. In the Middle East, a plasmid, designated by Anderson as F_1me, was very widely distributed in an *S. typhimurium* clone causing severe infections in pediatric units.[39] Inc HI2 plasmids were reported from 24% of the human isolates. Anderson[39] concluded that, in each region, ecological conditions had favored the emergence of particular strains of *S. typhimurium*, armed with particular plasmids, and that these strains had established themselves and spread and that they might maintain their predominance for long periods.[39] Anderson also noted that the outbreaks of human salmonellosis associated with these resistant *S. typhimurium* strains had affected pediatric units in particular and that a mortality rate of up to 30% was commonly associated with systemic invasion. Therapy in these cases were severely limited by the multiresistant nature (five or more drug determinants) of the dominant plasmid types.[39]

Anderson's report deals solely with the distribution of plasmid types in *Salmonella* serovars collected during the period 1968 to 1976. Subsequent monitoring of *Salmonella* R-factors in North America, Northern Europe, and Japan clearly indicates that the continued prevalence of Inc H types with time is unpredictable and highlights the difficulties involved in assessing the global distribution of individual plasmid types. Anderson failed to detect Inc H plasmids in 1276 *S. typhimurium* isolates from human and animal sources in France and Belgium prior to 1977. This was supported by the studies of Avril and co-workers on 410 strains belonging to six serotypes of epidemic *Salmonella* isolated mainly in France and Belgium between 1971 and 1975.[40] A very different Inc type prevalence was reported by Pohl and co-workers[36,41] from a study of 449 isolates from 21 different *Salmonella* serotypes from man and animals in France and Belgium between 1978 and 1981. Inc H plasmids were identified from 14 serotypes and constituted the dominant plasmid type, being present in 53% of the total isolates examined. The resistance pattern encoded by Inc H was predominantly CSSuT.[41]

By contrast, the prevalence of Inc H plasmids in *Salmonella* serotypes decreased in Canada between 1968 and 1980. Anderson reported that Inc HI2 predominated in 336 Canadian *S. typhimurium* isolates obtained from human and animal sources between 1968 and 1976.[39] The detailed studies of Taylor and co-workers on a variety of *Salmonella* serotypes isolated in Ontario from human and veterinary infections in 1974 confirmed the predominance of Inc HI2.[42-44] Grant and di Mambro initially investigated 589 human and 204 animal strains of *Salmonella* and showed that 12.4 and 38.2%, respectively, of the resistant human and animal strains carried self-transmissible plasmids, the majority of which encoded CKST resistance.[42] Of these plasmids, 12 were selected at random and all proved thermosensitive for transfer.[42] A more detailed investigation of the chloramphenicol-resistant strains by Taylor and Grant showed that resistance in most of the strains was mediated by HI2 plasmids.[43] A study of 40 chloramphenicol-sensitive *Salmonella* isolates from this collection showed that 7 of the 17 self-transmissible plasmids detected also belonged to the HI2 group.[44] Taylor et al. concluded that a HI2 plasmid which coded for CKST and which, on occasion, had lost the K or S genes, was extremely prevalent in *Salmonella* of a variety of serotypes in Canada.[44]

A subsequent study by Bezanson et al.[38] on *Salmonella* serovars from animal feeds, livestock sources, and from man in nine Canadian provinces in 1977 and 1978 showed that

Inc Iα plasmids occurred most frequently, seconded by Inc FII in animals and Inc H in humans.[38] Equal numbers of HI1 and HI2 plasmids were detected in animal *Salmonella* isolates whereas HI2 outnumbered HI1 by a factor of 2 to 1 in the isolates from human sources. All of the Inc H plasmids encoded resistance to CSuT. These studies suggested a decrease in the incidence of Inc H in Canadian *Salmonella* serotypes. This trend was verified by a subsequent study of 70 *Salmonella* strains belonging to 15 serotypes isolated from human sources in Ontario in 1979 and 1980.[45] Gel electrophoretic studies revealed 81 plasmids in a total of 26 of the resistant isolates — only one of which was of high molecular weight and proved on further study to belong to Inc HI2.

Anderson's study highlighted a particular prevalence of Inc HI1 plasmids in *S. typhimurium* isolates of human and animal origin from Southeast Asia (the isolates were mainly from Singapore and Malaysia).[39] Of the 180 conjugative plasmids analyzed, 67% belonged to the Inc HI1 group (Table 2). A significant variation during the time period studied was noted insofar as plasmids of the I_1 and FII groups were shown to predominate prior to 1972 whereas Inc HI1 types dominated between 1972 and 1976.[39] The latter dominance of Inc H was largely attributable to the clonal spread of an Inc HI1-carrying *S. typhimurium*, phage type 193. Subsequent studies by Japanese workers confirmed the prevalence of H plasmids in Asia, and in Japan in particular, and showed that they were not confined to a single *S. typhimurium* phage type but were present in a variety of *Salmonella* serovars and in other members of the Enterobacteriaceae.[37,46-49]

A study by Makino et al.[37] of R-plasmids from *S. typhimurium* of bovine origin, isolated from 1970 to 1979 in Japan, revealed a steady increase with time in the frequency of R-plasmid-carrying strains and demonstrated an overall incidence of 56.5% for plasmids of the Inc H group. More detailed typing indicated that 80% of the thermosensitive plasmids belonged to Inc HI1 with 20% Inc HI2.[37] The increasing prevalence of Inc H plasmids mediating multiple drug resistance was attributed by these authors to the widespread use of antibiotics in the calf-rearing industry in Japan.[37] Subsequent studies of *Salmonella* isolates from cattle, swine, and chickens confirmed the predominance of Inc H plasmids in Salmonellae of bovine origin (50%) and demonstrated a lesser, though significant, incidence of Inc HI1 plasmids in isolates of chicken and swine origin.[46] Ishiguro and co-workers carried out a more detailed investigation of *Salmonella* serotypes from swine in 1979 and 1980.[47] Transferable drug resistance was demonstrated for 69% of the isolates tested with Inc HI1 as the dominant self-transmissible plasmid type.[47] Yoshida and co-workers[48] also reported the prevalence of Inc HI1 and Inc HI2 plasmid types in *Salmonella* sp. isolated from clinical human sources in Japan. The widespread distribution of Inc H plasmids in Japan is further highlighted by the recent demonstration by Niida and co-workers[49] that both Inc HI1 and Inc HI2 plasmid types are carried by domestic and feral pigeons, crows, and kites in Japan and are present in *E. coli* isolates in addition to *Salmonella* sp.

It is apparent from the Japanese, Canadian, and European studies that the incidence of particular R-factor types can change quite dramatically in *Salmonella* sp. with time in a given geographical region. Bezanson and co-workers[38] and Makino et al.[37] suggested that changes in plasmid type distribution may simply reflect changes in antibiotic usage. It is likely, however, that a variety of other selective pressures and ecological/environmental factors are also involved.

B. Enterobacterial Species and Animal Host Distribution of Inc H Plasmids on a Global Scale

Inc H plasmids have been shown to occur in a wide variety of Enterobacterial species in association with a wide range of animal hosts. Because of the importance of *Salmonella* sp. in human medicine and in animal husbandry and because Inc H plasmids frequently specify chloramphenicol resistance (as well as commonly mediating multiresistance), the majority of studies carried out to date have been concerned with the incidence of H plasmids in *Salmonella* serotypes. In more recent years, the prevalence of Inc H in *E. coli* and in other members of the Enterobacteriaceae has been analyzed, with particular reference to the genetic relatedness of plasmids isolated simultaneously from *E. coli* and *Salmonella* from the same host animal or fecal source.

1. Incidence of H Plasmids in Salmonella

In *Salmonella typhi,* chloramphenicol resistance has been shown to be almost invariably associated with the carriage of Inc H plasmids which encode a variety of other antibiotic resistance determinants.[8,39] The majority of Inc H plasmids isolated to date from *S. typhi* belong to the HI1 group, although HI2 plasmids appear to be dominant in *S. typhi* isolates from Spain.[8,39] H plasmids are not confined to any single phage type of *S. typhi,* as evidenced by the isolation of nine distinct Vi types, carrying HI1 plasmids encoding CSSuT, CSSu, or ACSSuT resistances, during typhoid epidemics in South Vietnam and Thailand.[8]

S. typhimurium is the most ubiquitous of the Salmonellae associated with enteric disease in man and animals.[50] It is also the most predominant serovar among drug-resistant Salmonellae.[51] Both Inc HI1 and HI2 are widely disseminated in *S. typhimurium*, the predominance of either type varying geographically and with time as discussed above. In *S. typhimurium*, Inc H plasmids appear to be limited to a relatively small number of phage types,[50] e.g., types 193, 204, 204a, 204c, which have been implicated in clonal outbreaks of salmonellosis in calves[13-15] and in pediatric units.[38,39,43] Inc H plasmids are not confined, however, to *S. typhimurium* strains. Although plasmids of the H group were shown to predominate in *S. typhimurium* from clinical sources, Canadian studies also identified HI2 plasmids from other *Salmonella* serovars, e.g., *S. anatum, S. pensacola, S. typhimurium* var. *copenhagen.*[42,44] Pohl and co-workers[36] studied Inc H prevalence in isolates which were resistant to more than 3 drugs and which included 21 different *Salmonella* serovars from human and animal sources in France and Belgium.[36] Inc H plasmids were identified from 14 of the 21 serotypes tested although the prevalence of Inc H in different serotypes varied considerably.[36] Inc H was particularly prevalent in *S. schwarzengrund* (97% of isolates tested). Interestingly, H plasmids were not identified in any of the 169 multiresistant *S. dublin* isolates screened.[36] The basis for the observed prevalence of Inc H plasmids in some serovars and their absence from others is unclear and warrants further investigation.

With respect to host animal, Inc H plasmids in *S. typhimurium* and, to a lesser extent, in other *Salmonella* serovars, have been isolated from man,[37,39,41,45,48] from the common meat animals (cattle, pigs, and poultry),[15,36,37,46,47] from dogs, cats, and horses,[52] and from birds such as pigeons, crows, and kites.[49] Undoubtedly, the selective pressure exerted by the use of antibiotics for prophylactic, therapeutic, and growth promotion purposes in intensive animal raising is responsible for the high incidence of multiresistant R-plasmids in pigs, calves, and poultry. The demonstrated presence of H plasmids in wild, as well as in caged, healthy birds in Japan,[49] in the absence of any known selective pressure, suggests that Inc H plasmids are even more widely distributed in the animal kingdom than previously suspected. The source of Inc H plasmids to free-flying, feral birds in Japan is, as yet, unknown.

The highest incidence of salmonellosis in statutory animals (cattle, sheep, pigs, poultry) over the past 20 years in the U.K. has occurred in cattle and poultry.[50] Plasmid presence in the enterobacteria of calves is particularly high since the pressure for selection of antibiotic-resistant bacteria is greater in intensive calf rearing than in any other intensive livestock operation.[52] Newborn calves are particularly susceptible to enteric infection because they are generally colostrum deprived, are stressed and mixed with calves from other sources during transportation, and the crowded nature of their rearing facility favors feco-oral transfer of enteric pathogens. Antibiotics are generally administered in high dosage to combat both salmonellosis and colibacillosis caused by enteropathogenic *E. coli*.[52] Nevertheless, although Inc H-carrying serotypes of *Salmonella* have caused major epidemics in calf units in the U.K.[13-15] and in the U.S.,[53] Linton reported that many of the extensive outbreaks of calf salmonellosis in the U.K. over the past 20 years have been caused by fully sensitive strains of *S. typhimurium* and *S. dublin*.[50] Detailed studies of animal *Salmonella* isolates, which had been submitted to the U.K. Central Veterinary Laboratory between 1971 and 1979, indicated that, apart from sulfonamides and streptomycin, relatively few *Salmonella* isolates were resistant to a broad spectrum of therapeutically useful drugs.[50] Surveys in the Netherlands have yielded similar results.[54] The converse is true for *E. coli*, where a marked increase in the level of multiresistant strains has been noted over the past decade.[15,55]

However, it should be stressed that chloramphenicol resistance in animal *Salmonella* serotypes has increased dramatically in more recent years and has been often associated with Inc H plasmids encoding resistance to a number of other clinically important drugs. Epidemics caused by Inc H *Salmonella* serovars in intensive livestock units are usually of a clonal nature but may be distributed rapidly in a region because of animal transportation. Such outbreaks may result in major losses to the industry, as evidenced by the 50% mortality rate attributed to an Inc HI1 containing *S. typhimurium* in a calf unit in the U.S.[53]

The risk to man posed by the emergence of multiresistant Salmonellae in animals is also considerable. Human infection with *Salmonella* serotypes is undoubtedly of animal origin,[39] with poultry and pork being regarded as the most common sources.[56] *Salmonella* infections in humans in recent years have been largely associated with pediatric units[39,43,45] and have resulted in mortality rates of up to 30%, with systemic invasion and meningitis as common complications. The involvement of Inc H plasmids encoding resistance determinants to five or more of the clinically useful antibiotics in a significant number of these outbreaks has seriously compromised medical therapy. Although the original source of these Inc H *Salmonella* strains is undoubtedly animal, Anderson stresses the current significance of cycling in the human host, by nosocomial or other institutional spread.[39]

2. Incidence of Inc H Plasmids in other Enterobacteria

Although Inc H plasmids have been isolated from a variety of enterobacterial species, the data available is of a limited nature and does not permit any detailed conclusions as to the overall prevalence or geographical distribution of Inc H in enterobacterial species other than *Salmonella*. The majority of the studies to date have been concerned with the isolation of Inc H plasmids from *E. coli* from human and animal sources and from sewage and fecally contaminated waters.[15,49,51,57-62] A limited number of studies have confirmed the presence of Inc H plasmids in other enterobacterial species. HI2 plasmids were isolated from *Shigella flexneri* by Taylor and Grant.[63] Frost and Rowe[64] isolated six HI1 plasmids from a total of 323 strains of drug-resistant *S. flexneri* strains isolated in England and Wales between 1974 and 1978. Interestingly, four of these strains were from infections originating in India and two from infections in the U.K. Three additional HI1 plasmids, which were transfer defective, were found in *S. flexneri* strains isolated from Vietnamese immigrants.[64] Only one HI2 plasmid was detected from all the *Shigella* isolates examined and this too had been obtained from a Vietnamese immigrant.[64]

HI2 plasmids have been isolated from *Serratia marcescens* by Taylor and Grant[65] and by Hedges and co-workers.[22] These were initially classified as group S[22], but later correctly classified as belonging to the Inc HI2 subgroup.[20] All of the HI2 plasmids studied by Hedges and co-workers were from clinical isolates of *S. mercescens* obtained from hospitals in Boston and in Toulon, France. They were shown to mediate resistance to ampicillin, chloramphenicol, streptomycin, tetracycline, kanamycin, and gentamycin in varying combinations, with all of the gentamycin strains originating in the Toulon hospital.[22]

Inc HI2 plasmids were identified in two strains of *Citrobacter* isolated from sewage and sewage-contaminated river water by Smith and co-workers.[57] Taylor and Summers also identified Inc HI2 plasmids from *C. freundii*.[66] Romero and Perduca isolated chloramphenicol-resistant strains of *Enterobacter* from clinical specimens in Italy and showed by incompatibility testing that a small proportion of these strains harbored Inc S (i.e., HI2) R-factors.[61] Three HI2 plasmids were isolated by Smith and co-workers from *Klebsiella pneumoniae* strains from domestic animals in Great Britain.[57]

Richards and Datta also isolated a HI plasmid from two *K. aerogenes* isolates from a patient suffering from typhoid fever.[67] Although the subgrouping was not specified, the characteristics of the plasmids suggest that they belong to the Inc HI2 subgroup. Inc HII plasmids were also detected by Datta's group in the same *Klebsiella* isolates[67] and in a multiply resistant strain of *K. aerogenes* that infected at least 56 patients during a 5-month period in a London hospital.[21] The Inc HII plasmids were shown to be nonthermosensitive for transfer and to co-exist stably with the HI plasmid in the *K. aerogenes* strains for at least 40 generations at 37 or 42°C.[67]

In *E. coli*, plasmids belonging to both the HI1 and HI2 subgroups have been identified from human and animal sources in a variety of geographical regions. Linton and co-workers,[15] in a study of chloramphenicol resistance in *S. typhimurium* and *E. coli* in calves with endemic *Salmonella* infection, showed that 11% of the chloramphenicol-resistant *E. coli* isolates studied harbored Inc HI2 plasmids specifying from four to six drug resistance determinants. Since the incidence of chloramphenicol resistance in fecal *E. coli* in the calf unit varied from 15 to 75%, with an average incidence of 49% throughout the 34-day study period, the prevalence of Inc HI2 plasmids in this particular study was quite dramatic.[15] The HI2 plasmids were shown to occur in five O-serogroups and in nontypable strains of *E. coli*.[15]

By contrast, studies by Timoney on *S. typhimurium* and enteropathogenic *E. coli* isolates from calves in New York indicated a high incidence of HI2 plasmids in the *Salmonella* isolates (74% of transferring strains) whereas only one HI2 plasmid was identified from the 115 *E. coli* isolates investigated.[53] Ishiguro and co-workers studied the prevalence of Inc H plasmids in *Salmonella* serovars and in *E. coli* from a pig farm that had a long history of *Salmonella* infections.[51] Transferable drug resistance was demonstrated in 82% of the *Salmonella* isolates studied, with the majority of their plasmids being thermosensitive for transfer. Thermosensitivity, in all cases, was shown to be associated with Inc HI1 plasmids encoding a variety of multiple resistance determinants.[51] Thermosensitive plasmids were detected in 52% of the R$^+$ *E. coli* isolates and both HI1 and HI2 plasmid types were detected.[51]

Studies on Inc H prevalence in *E. coli* from sources other than intensive livestock units have generally revealed a low incidence of Inc H plasmids in both man and animals and in sewage and sewage-polluted waters. Smith and co-workers demonstrated the presence of both Inc HI1 and HI2 plasmids in *E. coli* isolates from human and animal specimens and from sewage.[57] The majority of the H plasmids identified belonged to the HI2 subgroup, although two HI1 plasmids, encoding tetracycline resistance, were identified from healthy pigs.[57] In a study of hospitalized and healthy patients in India, Rangnekar and co-workers identified Inc HI1 and HI2 plasmids in *E. coli* isolates of hospital origin only.[58] The overall incidence of HI1 and HI2 was markedly lower than that of other plasmid types, such as Inc H1, Inc C, Inc F$_1$*me*, etc.[58] Romero et al.[61] studied fi$^-$ R-factors specifying chloramphenicol resistance in clinical isolates of *E. coli*. Of the limited number of fi$^-$, Cm$^+$ *E. coli* isolates examined, 33% were shown to harbor Inc HI2 R-factors mediating CTKSSu or CSSu resistances. Thermosensitive R-factors have been studied more intensively in *E. coli* in Japan than elsewhere. Terakado and Sato[59] concluded that Inc HI plasmids frequently exist in naturally occurring strains of *E. coli* from cattle, pigs, and pigeons in Japan. More recently, Niida and co-workers[49] carried out a detailed study on plasmid types in *E. coli* isolates from feral and domestic pigeons, crows, and kites. A total of 63 R-plasmids were obtained from the 48 *E. coli* strains examined and 11.3% of these were classified as Inc HI1.[49] No Inc HI2 plasmids were detected in any of the *E. coli* isolates studied. Inc HI1 plasmids were also detected by Nakamura and co-workers[60] in *E. coli* isolates from mynahs imported into Japan from Thailand; 13% of 62 R-plasmids selected at random were shown to belong to the HI1 subgroup.[60]

It is clear from the above studies that Inc H plasmids occur in a wide range of enterobacterial species. Timoney and Linton concluded that, although HI2 plasmids exist in *Citrobacter, Serratia, Klebsiella, Shigella,* and *Escherichia,* they are relatively uncommon in these genera. This conclusion is supported by the studies of Smith and co-workers[57] on Inc H distribution in enterobacteria from sewage and polluted waters. Regional prevalences may exist, however, as suggested by the significantly higher incidence of Inc H in *E. coli* from animal sources in Japan.[49,60] These recent findings are in keeping with the known dominance of Inc H plasmids in *Salmonella* serovars from Japan and from Asia in general. The high incidence of Inc H, recorded by Linton and co-workers[15] in *E. coli* isolates from a calf unit in Britain, should not be regarded as a reflection of the overall prevalence of Inc H in *E. coli*. It simply reflects the abnormally high selective pressure imposed by antibiotic usage in calf-rearing facilities in general and the specific situation (endemic infection with Inc H containing *S. typhimurium*) prevailing on the farm at the time of sampling. Linton's[15] study highlights, however, the potential for serious public health problems in the future, should the indigenous *E. coli* of animals become carriers of H plasmids.

V. CONJUGAL TRANSFER OF Inc H PLASMIDS

Plasmids normally associated with the enteric bacteria exhibit optimum conjugal transfer efficiency at 37°C and transfer with markedly lower efficiency at temperatures below 37°C. The chief distinguishing feature of the HI group, therefore, is their transfer thermosensitivity, i.e., their capacity to conjugate most efficiently at temperatures below 37°C.[25,68] Screening for this trait has now been incorporated routinely as part of the preliminary characterization of plasmids of unknown incompatibility group.[69,70]

The genetic and biochemical basis of thermosensitive transfer is, as yet, poorly understood. In particular, the precise nature of the temperature-sensitive step in short-term mating studies has not been clearly defined. Few studies have been carried out, either, on ascertaining the optimal transfer temperatures for different Inc HI subgroups and in identifying underlying molecular differences.

Rodriguez-Lemoine and co-workers[71] studied the effect of the donor and recipient pre-incubation temperature on the transfer of S-plasmids (Inc HI2) at 22 and 37°C in 1-hr mating experiments. They concluded that the pre-incubation temperature of the donor was the major rate-determining factor and that neither the preincubation temperature of the recipient nor the temperature during mating were of primary importance. It appeared from these studies that prior growth of the donor at lower temperatures was necessary for the synthesis of a transfer system which could then function, irrespective of the mating temperature.[71] By contrast, later studies by Taylor and Levine[72] on HI1 and HI2 plasmids showed that both the growth temperature of the donor culture and the temperature of the mating mixture were rate determining. Transfer was not observed during a 2-hr mating period between 37 and 44°C even when the donor culture was pre-incubated at 22 or 26°C.[72] These authors[72] attributed the contrary findings of Rodriguez-Lemoine et al.[71] to failure to reach the inhibitory temperature during mating, due to the large mating volume and the short mating period utilized in their study. Taylor and Grant's findings were confirmed by Yoshida and co-workers with thermosensitive plasmids (presumably Inc H) from *S. typhimurium*.[73] They also found that the duration of the mating period was critical, i.e., the provision of a sufficiently long mating period at 22°C (> 2 hr) allowed synthesis of the transfer system, thereby masking any effect due to the pre-incubation temperature of the donor.[73]

Table 3
GROWTH RATES OF HI2⁺ AND HI2⁻
ISOGENIC *E. COLI* STRAINS AT 22
AND 37°C

Temperature (°C)	Generation time (min)		
	E. coli J53 Rif	*E. coli* J53 Rif pGU2[a]	*E. coli* J53 Rif pGU34[a]
22	120	120	120
37	29	28.5	28.7

[a]   pGU2 and pGU34 are Inc HI2 plasmids obtained, respectively, from *Enterobacter cloacae* and *E. coli* isolates from city sewage in Galway.

Transfer efficiency is also affected by the environment in which conjugation takes place.[74] Plasmids of the N and P groups, for example, transfer more efficiently on solid surfaces than in liquid media.[75,76] Bradley and co-workers[74] attributed this phenomenon to the type of pilus specified by the plasmid. Since Inc H plasmids determine flexible pili, they might be expected to transfer equally well in a liquid or on a solid surface. When mating was carried out on membrane filters[77] or on the solid surface of plates,[74] no significant increase in transferability was noted for plasmids of the HI group. By contrast, Yoshida and co-workers[78] found increased transfer frequencies for HI2 plasmids when mating was conducted on membrane filters rather than in broths. Variable results were obtained with plasmids of the HI1 subgroup.[78] Rodriguez-Lemoine and Cavazza[79] also demonstrated an increased transfer efficiency on solid surfaces and on membrane filters for HI2 plasmids in donor cultures grown under nonoptimal conditions.

Maher and Colleran studied the effect of temperature on donor growth rate and on HI2 plasmid maintenance.[62] As illustrated in Table 3, carriage of Inc HI2 plasmids did not appreciably affect the growth rate of a standard *E. coli* strain either at 22 or at 37°C. Growth at 37°C did not result in any detectable loss of HI2 plasmids from the standard *E. coli* hosts.[62] These studies confirm that thermosensitivity is solely associated with the plasmid transfer system and does not involve any additional plasmid effects on the host metabolism.

Mating frequencies at 22, 26, and 37°C for standard HI1, HI2, and HI3 plasmids[72] and for HI2 plasmid isolates from sewage enterobacteria in Galway[62] are presented in Table 4. Clearly, transfer at temperatures between 22 and 26°C permits considerably higher mating frequencies than at 37°C for plasmids from all the HI subgroups. Transfer thermosensitivity is not characteristic of HII plasmids.[21] Whether an optimum temperature for plasmid transfer exists within the HI group or HI subgroups has not yet been clearly ascertained. The majority of workers have described H plasmid thermosensitivity as the ability to transfer more efficiently at arbitrary temperatures between 22 and 30°C than at 37°C. Smith and co-workers, in a study of the transferability of HI plasmids from R⁺ enterobacteria to *E. coli* K12, categorized the plasmids into two sets which displayed different temperature optima for conjugation.[57] Plasmids in Set I exhibited optimal transmissibility at 22°C whereas Set II plasmids transferred efficiently over the temperature range 22 to 33°C. Plasmids of the HI2 subgroup were found in both sets whereas HI1 plasmids were observed only in Set II.

Transmissibility was poor at 37°C for virtually all the plasmids studied, with the exception of some HI2 isolates.[57] Frost and co-workers considered the ability to transfer, albeit poorly, at 37°C as being a well-defined trait of differentiation between the HI1 and HI2 subgroups, the HI1 group being regarded as incapable of transfer at 37°C.[70] The duration of the mating

Table 4
MATING FREQUENICES FOR PLASMIDS OF THE HI1, HI2, AND HI3 SUBGROUPS AT VARYING TEMPERATURES

Plasmid	Inc group	Mating frequency[a]			Ratio of mating frequency	
		22°C	26°C[b]	37°C	(22/37°C)	(26/37°C)[b]
pRG1251	HI1		1×10^{-3}	$<1 \times 10^{-8}$		$>1 \times 10^{5}$
R477-1	HI2		2×10^{-3}	$<1 \times 10^{-8}$		$>2 \times 10^{5}$
R478	HI2		9×10^{-3}	$<1 \times 10^{-8}$		$>9 \times 10^{5}$
R828	HI2		6×10^{-5}	$<1 \times 10^{-8}$		$>6 \times 10^{3}$
M1P233	HI3		2×10^{-3}	$<2 \times 10^{-6}$		$>1 \times 10^{3}$
pGU2	HI2	1.6×10^{-3}		$<1 \times 10^{-7}$	$>1.6 \times 10^{4}$	
pGU3	HI2	2×10^{-3}		$<1 \times 10^{-7}$	$>1.6 \times 10^{4}$	
pGU34	HI2	1×10^{-2}		$<1 \times 10^{-7}$	$>1 \times 10^{5}$	

[a]   Calculated as the number of transconjugants per number of recipients after overnight mating. Ratio of 1:5 for donor and recipient cells.
[b]   Data from Taylor and Levine.[25]

period must, however, be clearly defined since R27 (TP117) — a standard HI1 plasmid — was found to transfer after 24 hr but not after a 2-hr mating period.[9]

VI. MECHANISM OF CONJUGAL TRANSFER

The mechanism of conjugation is poorly understood for most enterobacterial plasmids, including those of the Inc H group. Detailed analysis to date has been confined mainly to studies of the F-factor and F-like plasmids. The F mating process can be divided into three stages: (1) collision leading to effective pair formation; (2) transfer of the DNA from donor to recipient cell, and (3) processing of the DNA in the recipient cell.[80] Complementation analysis of F transfer mutants has identified transfer genes which have been subsequently mapped and ordered using deletion mutants, insertion mutants, and cloned transfer segments in phage vectors and nonconjugative plasmids. Polar mutations have also facilitated understanding of the F transfer genes which are grouped together in one large operon.

These and other physical and biochemical studies have permitted determination of the general role of the F transfer genes. However, little is yet known biochemically of the enzymic functions of transfer gene products involved in pilus formation, in the physical interaction between donor and recipient cells, in conjugal DNA metabolism, or in the regulation of the transfer process.[81,82]

The lack of interaction between conjugative systems encoded by plasmids from different Inc groups, allied to morphological and serological differences in their pili[74] and an overall lack of homology between plasmid DNA from different Inc groups, suggests the existence of numerous distinct conjugation systems. Several transfer systems have now been located and mapped. As with the F-factor, the transfer genes of the Inc N plasmid, R46, are located within a single operon.[83] Studies on the Inc P plasmids, R91-5 and RP4, have identified two and three transfer regions, respectively,[84-86] and at least two transfer segments have also been detected for the Ti plasmid, pTiC58.[87] The first genetic and molecular analysis of a H plasmid transfer system was reported by Taylor[88] following the successful isolation of usable quantities of H plasmid DNA.[89] Transfer analysis was undertaken for the HI1 plasmid R27 (TP117) following Tn7 mutagenesis to obtain tra⁻ mutants.[88] Changes in fragment size, coupled with hybridizations with radiolabeled Tn7 probe, identified DNA restriction fragments involved in transfer. Concomitant work in the same laboratory to produce the first

Table 5
MOLECULAR WEIGHTS OF RESISTANCE PLASMIDS OF ENTERIC BACTERIA

Inc group	Plasmid	Resistance pattern[a]	Molecular weight (megadaltons)	Ref.
HI1	R27	T	112	90
HI1	R726	CSSuT	120	90
HI1	Tp124	CSSuT	120	91
HI2	R478	CKTHgTe	166	65
HI2	Tp116	CSSu	143	91
HI2	pWR23	Suc⁺lac⁺Te	164	92
HI3	M1P233	Suc⁺Te	150	24
HII	pHH1507a	Te	90	67
HII	pHH1508a	STpTe	100	21
HII	pHH1457	ACSSuKTGTpHgTe	110	21
FI	R386	T	74	91
FII	R1	ACSSuK	62	91
K₁	R144	TK	62	91
W	R388	SuTp	21	91
N	N3	SSuT	33	91
P	RP4	AKT	36	91
Q	R300B	SSu	5.7	91

[a] Ampicillin (A), chloramphenicol (C), streptomycin (S), sulfonamides (Su), kanamycin (K), gentamycin (G), tetracycline (T), trimethoprim (Tp), mercury (Hg), tellurite (Te), sucrose utilization (Suc⁺), lactose utilization (lac⁺).

restriction map of a H plasmid, together with a more detailed analysis of the tra region by Tn5 and Tn7 insertion, has identified at least two separate regions involved in transfer.[173]

No specific information is, as yet, available on the transfer systems of other H plasmids although it is known that there is little homology between the DNA of plasmids from different H groups and subgroups. Among F-like plasmids, the products of the pilus-synthesizing genes are interchangeable, while other conjugative gene products are often specific to plasmids within a particular F incompatibility group and do not cross-react. A similar degree of specificity may exist among H plasmids belonging to different groups or subgroups.

VII. MOLECULAR SIZE OF Inc H PLASMIDS

A key distinguishing characteristic of Inc H plasmids is their large molecular size. Unlike other well-characterized enterobacterial plasmid types, the molecular weight of Inc H plasmids is typically in excess of 100 Mdaltons and may be as large as 200 Mdaltons. The smallest naturally occurring Inc H plasmid so far isolated has a molecular size of 90 Mdaltons and belongs to the Inc HII group.[67] The molecular size of representative members of the Inc H groups is compared with that of other enterobacterial plasmid types in Table 5. It is apparent from Table 5 that HI2 plasmids tend to be larger than plasmids of the other HI subgroups or of group HII.

VIII. ISOLATION OF H PLASMID DNA

Purified plasmid DNA is required for genetic studies involving the subcloning of individual plasmid DNA fragments, for transformation experiments, and for use as a radioactive probe.

Isolation of small plasmid DNA is now a simple and routine laboratory procedure, the major objectives being to avoid plasmid breakage and plasmid loss both during and after cell lysis and to ensure that the plasmid DNA is free from contamination by chromosomal DNA. Since the physical properties of large plasmids resemble, more closely than for smaller plasmids, the properties of chromosomal DNA, purification of sufficient quantities of large plasmid DNA is difficult. Extensive plasmid loss tends to occur due to shearing and coprecipitation with chromosome-membrane complexes. Several methods have been developed specifically for large plasmid isolation.[93-98] The majority of these procedures take advantage of the resistance of plasmid DNA molecules to strand separation at alkaline pH (12 to 12.5). Following neutralization, chromosome-membrane complexes are separated from the plasmid DNA by salt precipitation,[94,96,97] phenol extraction,[95] or centrifugation.[93,98] The manipulations are performed as gently as possible in order to avoid extensive shearing.

Because of their exceptionally large size, isolation of Inc H plasmid DNA poses particular difficulties. In our laboratory, Inc H plasmids were subjected to a variety of alkaline denaturation isolation procedures with limited success. Yields, in general, were poor and separation from chromosomal DNA proved difficult. These findings are in agreement with the poor yields reported by Humphreys and co-workers[93] and by Hansen and Olsen.[94]

Using a modified version of the alkaline denaturation technique of Currier and Nester,[95] H plasmid DNA was purified and restricted with Eco R1.[41] More recently, high yields of H plasmid DNA, of sufficient purity to permit restriction and hybridization studies, have been obtained using a neutral sarkosyl lysis procedure followed by separation of plasmid DNA from chromosomal DNA solely by dye buoyant density gradient centrifugation.[89] This procedure has proved successful in our hands and initial studies suggest that the yield may be enhanced by the use of large guage needles (Monoject 200: Sherwood Industries) for extraction of the plasmid bands from the gradients.[174]

IX. DNA HOMOLOGY STUDIES

Molecular hybridization of nucleic acids has been used by many workers to establish relatedness between plasmid deoxyribonucleic acids (DNAs) and, in particular, to determine whether any correlation exists with the classification obtained from genetic studies. A high degree of reassociation is normally observed between single-stranded DNA obtained from plasmids within the same incompatibility group.

Early studies by Grindley and co-workers,[99] prior to the subdivision of the HI plasmids into subgroups, showed that one of the four H plasmids under study did not reassociate with the others. Anderson[100] subsequently confirmed that there was little homology between plasmids belonging to the H1 (HI1) and H2 (HI2) subgroups. Hybridization studies on nitrocellulose filters by Roussel and Chabbert[24] again emphasized the lack of homology between HI1 and HI2 plasmids and led to the classification of a third subgroup, HI3, containing a single Inc H plasmid, M1P233, which was strongly incompatible with all HI1 and HI2 plasmids tested.

The availability of pure H plasmid DNA has recently allowed characterization of representative plasmids from the three HI subgroups by restriction enzyme digestion and southern transfer hybridization.[89] Although a diversity of restriction enzyme cleavage patterns was noted among the HI subgroups, similar cleavage patterns were observed within a subgroup. This was particularly true of the HI1 plasmids, whereas HI2 plasmids gave more diverse patterns. When the HI1 plasmids were probed with a nick-translated HI1 plasmid, extensive sequence homology was observed. A small amount of homology was noted between the HI1 probe and plasmids of the HI2 and HI3 subgroups and a representative plasmid, pHH1508a, from the HII group.[89]

X. H PILI

Conjugative pili of variable morphology are determined by representative plasmids from all incompatibility groups in a standard *E. coli* host, such as *E. coli* K12. Within an incompatibility group, plasmids determine morphologically similar and serologically related pili. Three basic types have been identified — thin flexible, thick flexible, and rigid filaments.[101]

Morphological studies by Bradley[102] led to the original classification of HI1 and HI2 pili as rigid. On re-examination by immune electron microscopy, the pili specified by both HI subgroups were clearly shown to be thick and flexible.[101] Pili of the HI1 and HI2 subgroups were shown to be serologically related — a surprising finding because of the known lack of homology between HI1 and HI2 plasmid DNA.[24,89]

Production of H pili has also been observed for plasmids of the HII group.[21] These nonthermosensitive plasmids determine constitutive production of H pili[21] unlike the repressed synthesis reported for HI1 and HI2 subgroups.[26,101] The unique pili of the HI3 subgroup (as demonstrated by Bradley[175] for plasmid MIP233) appear to be of the rigid type but are atypical in being very short (only 125 nm long) and in mediating both surface and liquid conjugation — a characteristic normally associated with flexible pili. The small amount of DNA homology between HI1, HI2, and HII plasmids, as revealed by the restriction digest and probe analysis studies of Whitely and Taylor,[89] may constitute, in part, genes involved in H pilus production.

XI. PHAGE INHIBITION STUDIES

The ability of R-factors to inhibit the normal development of double-stranded DNA phage has been known for many years.[103,104] This phenomenon is normally referred to as bacteriophage inhibition (Phi) and it involves the restriction of incoming phage DNA by a restriction enzyme (host DNA is protected from the enzyme by a modification system). As a result, the efficiency of plating (EOP) phage is reduced and those phage that develop have their DNA modified. The EOP of such modified phage increases upon reinfection of the R^+ host.

Using three different bacterial hosts, Taylor and Grant[43] found that HI2, but not HI1, plasmids, expressed a "phage reduction" effect. Reduced EOPs were obtained for λ, T1, T7, and the lambdoid phages Ø80 and Ø81 in the presence of HI2 plasmids. No inhibition was noted for P1 and T4 phage. The phenomenon involved was different from the usual restriction and modification effect encountered with both fi^+ and fi^- factors[105] since lysates of inhibited phage did not overcome restriction.[43] Further studies revealed that, among 14 Inc groups tested, only HI2 plasmids mediated a pattern of inhibition to λ, T1, T5, and T7.[63] Plasmids from other incompatibility groups mediate different patterns of phage inhibition.

Although λ, T1, T5, and T7 inhibition is a pattern associated with all HI2 plasmids, the degree of inhibition may vary with both the host[63] and the individual plasmid[44] utilized. A lower level of inhibition to λ and T5 was shown to be mediated by HI2 plasmids from *Salmonella* sp. isolated in Ontario than by HI2 plasmids from other sources.[44] This lower level of inhibition has been termed "intermediate phage inhibition". Irrespective of whether a H plasmid mediates high or intermediate levels of reduction in EOP, the surface exclusion properties of all Inc H plasmids examined have been shown to be identical.[92] Since high level inhibition may be reduced to intermediate level inhibition by ethyl methyl sulfonate (EMS), the difference between the two levels may result from a single-step mutation.[44]

XII. STABILITY OF H PLASMIDS

Inc H plasmids are, in general, stably maintained in *E. coli* K12 at 37°C, suggesting that vegetative replication of the plasmids is not thermosensitive. Maintenance of H plasmids appears to be less efficient in *Salmonella* hosts. Humphreys[106] showed that a HI1 plasmid dissociated in *S. typhimurium* phage type 36 to yield a small R-determinant factor and a large tra[+] factor. Rangnekar[107] studied the maintenance of HI plasmids in *S. typhi* under different environmental conditions, i.e., in the presence of bile salts, at pH values from 6.3 to 7.3, and under anaerobic conditions. Both HI1 and HI2 plasmids were lost from the *S. typhi* host at rates which varied up to 20%. Taylor's group also reported that H[+] *Salmonella* strains lost their resistance phenotypes during storage at room temperature for 2 to 3 months.[42,44]

The stability of HII plasmids has been investigated by Datta and co-workers[21] using two HII plasmids originally obtained from *K. aerogenes* isolates. Plasmid pHH1457 was found to be unstable in both *Klebsiella* and *E. coli* K-12, being lost completely from the host at a rate of 0.75% per generation in nutrient broth. The plasmid also gave rise to variants which lacked 25 Mdaltons of DNA, most of the original drug resistances and were nonconjugative and determined no pili.[21] By contrast, another HII plasmid, pHH1508a, was shown to be stably maintained in both hosts.

High-temperature elimination studies by a variety of workers yielded variable results for HI plasmids. Rodriguez-Lemoine and co-workers found that Inc S (HI2) plasmids, which were originally detected in *Serratia marcescens*, were stable in *E. coli* K12 during growth at 30 and at 43°C. Williams Smith[57] studied the stability of 73 thermosensitive plasmids, which had been isolated from a variety of enterobacterial species and transferred to *E. coli* K12; 34 of the plasmids were stably maintained after three daily passages at 44°C in broth culture (plasmid loss was determined solely by loss of drug resistance determinants).[57] The rate of loss of the remaining 39 plasmids varied from 1 to 100%. In general, HI1 plasmids appeared to be more stably maintained than plasmids of the HI2 subgroup during growth at 44°C.[57]

By contrast, an Inc HI1 plasmid, which was harbored by three different hosts (*E. coli* K12, *E. coli* C, and *S. typhimurium*), was shown by Taylor and co-workers to display a marked level of instability as determined by drug resistance loss and gel analysis.[25] Furthermore, H plasmids from a common origin were found to share similar stability phenotypes, suggesting that the trait is dependent both upon the original bacterial host and the geographical background from which the plasmid was derived.[25]

XIII. CHARACTERISTICS ENCODED BY Inc H PLASMIDS

Because of their clinical and veterinary significance, the majority of studies on the characteristics encoded by Inc H plasmids have been concerned with drug resistance determinants. The exceptionally large size of Inc H plasmids suggests, however, that a variety of other phenotypic characters may be encoded by H plasmid DNA. Studies to date have revealed determinants for sugar and tricarboxylic acid utilization, for heavy metal resistance and for colicin B resistance.

A. Sugar Utilization

Screening for the range of carbohydrates metabolized by an enterobacterial isolate is one of the key biochemical tests employed in diagnostic bacteriology. The isolation of organisms which atypically utilize one or more carbohydrates obviously compromises the validity of these tests. The genes which code for individual sugar utilization have been shown to be encoded by plasmids from a range of different incompatibility groups. Inc H plasmids which carry the genes for lactose and/or sucrose utilization have been isolated from a number of different *Salmonella* serovars. Timoney and co-workers reported the presence of a lac[+] HI1 plasmid in a highly virulent strain of *S. typhimurium*.[108] Lac[+] HI2 plasmids were also isolated from *S. oranienberg*[23] and *S. infantis*.[109] An Inc HI2 plasmid, which was isolated from *S. tennessee,* was shown to encode both sucrose and lactose utilization genes.[110] Thermosensitive plasmids (presumably Inc H) from *S. thompson* were also shown to mediate sucrose utilization.[35] The sole plasmid which has been assigned to the Inc HI3 subgroup encodes genes for sucrose utilization.[23]

B. Citrate Utilization

The utilization of citrate as sole carbon source is an atypical trait of *E. coli*. It may be mediated either by a chromosomal mutation[111] or by the carriage of a plasmid encoding citrate-metabolizing genes. Early Japanese studies observed an association between citrate utilization and conjugal transfer thermosensitivity.[112,113] Williams Smith and co-worker[57,114] showed that thermosensitive plasmids of the HI1 subgroup could transfer the cit phenotype from their original *S. typhi* host to *E. coli* K12. Subsequently, other workers reported on the isolation of HI1 plasmids encoding the cit determinant from *E. coli* of animal and bird origin[49,115,116] and from *Salmonella* strains of bovine origin.[117,118] The cit determinant is not universally encoded by HI1 plasmids,[37,49,119] nor is it confined to plasmids of the H incompatibility group, having been shown to be encoded by Inc W and by other untypable plasmids.[115-117,120,121]

A range of other tricarboxylic acids, including isocitrate, *cis*-aconitate, *trans*-aconitate, and tricarballylate can also be utilized as sole carbon source by *E. coli* strains which harbor cit$^+$ R-factors. Ishiguro and co-workers[116,117] examined the tricarboxylic acid utilization capabilities of a range of cit$^+$ R-factor-containing *E. coli* isolates from animal and bird sources. Three distinct utilization patterns were noted, all three being unique to *E. coli* and none being associated exclusively with any one of the Inc groups under study.[116,117]

Recent studies indicate that protons (H$^+$ ions) are required for efficient cotransportation of citrate across the *E. coli* membrane.[122] Cations other than H$^+$ appear to function as cotransporters for citrate in other enterobacterial species. The difference in cation requirement, coupled with the unique tricarboxylic acid utilization patterns encoded by *E. coli* cit$^+$ R-factors, suggests a different origin for the cit character in *E. coli*.

Studies have also been carried out in order to determine whether the cit determinant is transposable.[123] None of the HI1 plasmids examined were capable of transposing their cit determinant to a phage although a cit transposon was detected on an untypable plasmid.[123] However, the involvement of cit transposons in HI1 plasmids cannot be ruled out by these experiments since transposons of larger size than the DNA packaging capacity of the λ phage would not have been detected by the technique used.[123]

Hybridization studies between HI1 plasmids and a cloned cit determinant from an Inc W plasmid showed no DNA homology between the cit genes carried by plasmids from these two Inc groups.[124]

C. Metal Resistance

Heavy metals from urban and industrial wastes are increasingly accumulating in rivers, lakes, soil, and sediments. There is considerable current interest in the interactions between bacteria and heavy metals because of the significance of biological transformation of metals in the environment. Enterobacterial plasmids are known which encode resistance to a variety of metals, including arsenate (As), silver (Ag), cobalt (Co), mercury (Hg), nickel (Ni), and tellurium (Te). Several reports indicate that metal ion resistance occurs more frequently in natural populations of Gram-negative bacteria than hitherto suspected.[125]

Resistance to As, Hg, and Te has been shown to be associated with plasmids of the Inc H group. Metal resistance is particularly characteristic of HI2 plasmids, with Te resistance being most commonly encountered. The single plasmid belonging to subgroup HI3 and the small number of plasmids assigned to the HII group also encode Te resistance. Hg resistance is more widely distributed, having been reported also for HI1 plasmids.[37,119]

1. Hg Resistance

Among the enterobacteria, plasmid-encoded resistance to Hg and organomercurials may be divided into two classes based on the spectrum of resistance conferred: (1) narrow spectrum in which resistance to Hg alone is mediated by a reductase which volatilizes Hg^{2+} to Hg^{o}[126] and (2) broad spectrum in which resistance is also conferred to phenylmercury and thiomersal by a different enzyme, organomercurial lyase.[126,127]

Plasmids belonging to Inc S (now HI2), Inc A-C, and Inc L (now Inc M) confer broad spectrum resistance. Inc A-C plasmids are atypical in being sensitive to thiomersal.[128] Although HI1, HI3, and HII plasmids which encode mercury resistance are known, no information is, as yet, available on the resistance spectrum or enzymic function specified by these H plasmid groups. Broad spectrum resistance to Hg and organomercurials is also encoded by Mer plasmids of *Staphylococcus aureus*[129] and by some P2-P8 plasmids of *Pseudomonas aeruginosa*,[130] but both the spectrum of resistance and the enzymic mechanisms are dissimilar to those of the HI2 group.

An initial study on the transposability of the HI2 Hg determinant indicated that a transposition-like mechanism was involved.[131] However, the variable amount of DNA transposed, the absence of inverted repeats and the inability of the segments to retranspose casts doubt on the involvement of a singular, unique transposable element.

2. As Resistance

Analysis of a collection of Hg-resistant isolates from hospital sources revealed that arsenate resistance (asa$^+$) was encoded by thermosensitive plasmids of the Inc S (HI2) group.[128] As resistance was also detected during a study of Te resistance among HI2 plasmids by Summers and co-workers.[132] Williams Smith[57] showed that the asa determinant was encoded by HI2 plasmids which had been obtained from enterobacteria from a number of different environments. As resistance was not associated with any of the HI1 plasmids investigated.[57]

The biochemistry of As resistance has been investigated using the FI plasmid, pR773, and asa$^+$ plasmids from *Staphylococcus aureus*.[133] Both systems were found to be inducible and to be biochemically indistinguishable. As is normally brought into the cell by competition with phosphate for the phosphate transport system in the membrane. The presence of an asa$^+$ plasmid does not alter the kinetic constants of the transport system nor does it detoxify arsenate.[133] Instead, arsenate is rapidly excreted from asa$^+$ cells, probably via an ATP-dependent efflux system.[134,135]

3. Te Resistance

Resistance to Te is frequently mediated by HI2 plasmids although it is not an invariant characteristic of this subgroup nor is it confined to plasmids of the H incompatibility group.[66,132] Neither the biochemical mechanism nor the genetic basis of Te resistance is clearly understood. Studies carried out to determine whether the resistance is carried by a transposable element have yielded negative results.[67]

D. Resistance to Colicin B

Bacteria often produce substances that inhibit or kill closely related species. These agents are known as bacteriocins and constitute a diverse group of substances that are usually proteinaceous in nature. Bacteriocins produced by *E. coli* are known as colicins.

E. coli K12 strains are normally sensitive to the action of colicin B. The presence of a HI2 plasmid has been reported to specifically protect the host *E. coli* K12 cell against the lethal action of colicin B.[136] This characteristic, termed Pac B, is unique to the HI2 subgroup but little is known, as yet, about the mechanism involved.[136]

XIV. ANTIBIOTIC RESISTANCE DETERMINANTS ENCODED BY Inc H PLASMIDS

Inc H plasmids generally confer simultaneous resistance to a variety of anitmicrobial drugs. As is evident from the limited number of H plasmids illustrated in Table 5, resistance determinants for most of the antibiotics used in clinical practice have been shown to be encoded by plasmids of the various H subgroups.

The relative incidence of different combinations of drug resistance determinants within the H incompatibility group is difficult to assess. Anderson found that the CSSuT profile was commonly associated with the HI1 subgroup[8] in salmonellae as did Linton and co-workers for HI1 plasmids in *E. coli* isolates of calf origin.[15] CSSuT resistance was, however, also mediated by other H plasmids and by plasmids from other Inc groups.[8] In Rangnekar's study, ACSSuT was the dominant Inc H profile encountered and the pattern was also shown to be prevalent in other Inc groups in the geographical region under study.[58] Clearly, the dominance of a particular antibiotic resistance profile within an Inc H group or subgroup largely reflects the current usage of antibiotics in a given area and at a given time and should not be considered as a universal or invariant profile for any particular plasmid group or subgroup.

Although chloramphenicol resistance may be mediated by plasmids from the FI, FII, I, or M incompatibility groups, it was initially assumed that the chloramphenicol determinant was an integral coding region on Inc H plasmids. Studies by Taylor's group showed that CSKT was the dominant resistance profile in HI2 plasmids in Canada and Taylor concluded that Inc HI2 plasmids which did not mediate chloramphenicol resistance were derived by spontaneous mutation of this central pattern.[44] Studies by Williams Smith,[57] Terakado et al.,[46] and Datta and co-workers[21] showed that H plasmids which did not encode chloramphenicol resistance could be isolated from a variety of enterobacterial species. Retrospective studies by Anderson also showed that isolates of *S. typhimurium* phage-type 193, which had been obtained between 1971 and 1973, harbored HI1 plasmids mediating either AK or AKT resistance.[8] It is now generally accepted that the chloramphenicol resistance determinant is not an integral coding region on Inc H plasmids, although it is one of the most widely encountered resistances specified by plasmids of the H group.

High-level resistance to chloramphenicol is usually mediated by chloramphenicol acetyl-transferase (CAT), an intracellular tetrameric enzyme which acetylates and inactivates the antibiotic using acetyl-CoA as acyl donor.[137-139] At least 15 naturally occurring variants of CAT are known and these can be differentiated from one another by several criteria, including plasmid linkage, mode of synthesis, molecular weight, native electrophoretic mobility, thiol reagent sensitivity, affinity chromatography, and immunological class.[140]

R-factors are known to determine three different CAT enzymes which can be distinguished by MIC values, K_m values for chloramphenicol, electrophoretic mobility, and inhibition by DTNB.[141] Type I CAT is commonly encoded by fi$^-$ plasmids from a variety of incompatibility groups, including the HI2 subgroup.[142] Maher and Colleran[143] showed that Type I CAT was also specified by HI2 plasmids from different enterobacterial origins and concluded that the Type I enzyme is conserved within the subgroup. The HI2 plasmids investigated were also shown to mediate fusidic acid resistance. This is in agreement with previous findings that the Type I CAT enzyme also mediates fusidic acid resistance due to the strong affinity of the enzyme for this steroid antibiotic.[144]

The chloramphenicol resistance genes on some R-factors are known to be encoded by Tn9 or one of the well-characterized derivatives of the Tn9 transposon.[145] Terakado and co-workers[146] showed that chloramphenicol resistance was carried by Tn3351, a Tn9-like transposon, on a plasmid belonging to the Inc H group.

Studies with other enterobacterial plasmid types have identified a large number of transposons which carry the genes for resistance to one or other of the commonly used antibiotics. Transposons which transpose at high frequency are generally widely distributed among plasmids of many groups whereas those which transpose at low frequency are found in a narrower range of plasmid types.[91] Few studies to date have examined the transposability of the drug resistance genes carried by H plasmids. Richards and Datta[67] confirmed the transposability of the resistance determinants for trimethoprim and streptomycin/spectino-mycin, which were carried by a HII plasmid isolated from a *K. aerogenes* strain obtained from a patient suffering from typhoid fever. Digestion of the plasmid DNA with Hind III, followed by gel separation and screening for the characteristic internal fragments of Tn7, indicated that the HII plasmid carried a Tn7-like transposon which determined a type I dihydrofolate reductase.[67] Interestingly, investigation of a small untypable plasmid from Tp[+] *Salmonella typhi* isolates from the same patient showed that it too carried a Tn7-like transposon.

HI2 plasmids from a variety of geographical areas and bacterial hosts are currently under study in Galway with respect to the transposability of their resistance genes. The molecular relatedness of any transposon encountered to well-characterized transposons such as Tn7, Tn9, etc. is being investigated in an attempt to obtain a better understanding of the origin and evolution of multiresistant plasmids of the Inc H group.

XV. VIRULENCE

The involvement of H plasmid-carrying strains of *S. typhi* with major epidemics of typhoid fever[8,39,147] and of H[+] *S. typhimurium* and other *Salmonella* serovars with serious outbreaks of salmonellosis in pediatric units[38,39] and in intensive livestock-rearing facilities[13-15] has prompted consideration of the possibility that H plasmids contribute either to the virulence or to the colonization capabilities of the host bacterium.[148]

The determination of a specific virulence toxin by plasmids of either the HI1 or HI2 subgroups has, so far, not been demonstrated. An increase in pathogenicity was not observed after oral administration of *S. typhimurium* strains harbouring HI1 or HI2 plasmids to chickens[57] or after subcutaneous or intraperitoneal injection of mice.[53,57,149]

A study of HI2 plasmids from *E. coli* isolates by Maher and Colleran[62] failed to detect either heat-stable or heat-labile enterotoxins which were specifically encoded by the H plasmids. Taylor and co-workers[173] could not determine any molecular relationship between the high molecular weight enteroinvasive plasmids found in pathogenic strains of *E. coli* and representative Inc H plasmids.

It is not clear either whether an association exists between virulence and *Salmonella* phage type. The Vi phage type of an *S. typhi* strain from the bowel of a patient suffering from enteric fever was shown to be altered from phage type A to degraded Vi by acquisition of an Inc H plasmid.[150] The phage type of the strain responsible for the severe Mexican typhoid epidemic in 1972 to 1974 was degraded Vi, suggesting a possible association between the acquisition of the HI1 plasmid and the exceptional severity and protracted nature of the Mexican typhoid fever.[149] However, the HI1-containing *S. typhi* strains implicated in the typhoid epidemics of Southeast Asia belonged to nine different Vi phage types[8] and studies by Threlfall[13] showed that the acquisition of HI1 or HI2 plasmids by epidemic strains of *S. typhimurium* did not alter the phage type.

Few studies have been initiated to determine whether H plasmids enhance the stability and colonization properties of their bacterial hosts. Butler and co-workers[149] showed that the multiplication of an *S. typhimurium* strain in the spleen of a mouse was not altered by the acquisition of a HI1 plasmid. Timoney and Linton[151] studied the persistence of an *E. coli* strain carrying the HI2 plasmid, pJT4, in the calf intestine. Parallel experiments with plasmid negative, isogenic strains clearly showed that pJT4 carried genes which enhanced the ability of the host strain to maintain itself in the intestine. The enhanced persistence did not appear to be related to the presence of colonization antigens and Timoney and Linton[151] speculated that pJT4 may code for changes in the outer membrane that lessen immunogenicity or that enhance resistance to gut immune mechanisms.

Plasmids which do not confer any form of virulence on their host cells may augment the success and communicability of the bacterium under specific conditions. A particularly clear example of this phenomenon is provided by a salmonellosis outbreak in a calf unit in New York.[53] The *S. typhimurium* strain involved carried a lac⁺ CSSuT plasmid belonging to the HI1 subgroup.[108] The ability to utilize the lactose in the milk replacer diet used in the veal unit apparently allowed the *S. typhimurium* strain to rapidly attain the lethal numbers which were responsible for the observed 50% mortality rate.[53] The combination of lactose positivity and multiresistance obviously presented a formidable problem in respect of therapy, diagnosis, and pathogenicity.

XVI. R-FACTOR EPIDEMIOLOGY

Clearly, the therapeutic use of antibiotics in human and veterinary medicine, coupled with their administration for prophylactic and growth promotion purposes in intensive livestock units, has contributed greatly to the current widespread distribution of R-factors in commensal and pathogenic bacteria. In considering the epidemiology of R-factors and the public health significance of the increased incidence of R-factors in the enterobacteria in particular, the following factors merit consideration: (1) the incidence of R-factors in animals and their onward transmission to man, (2) in vivo transfer of R-factors between commensal bacteria and between commensal and pathogenic bacteria in animals and man, (3) conjugal transfer of R-factors outside the body in feces, sewage treatment plants, fecally contaminated foods, water, or sediments, (4) the ability of R-factor-carrying strains of animal origin to colonize the human intestine, (5) conjugal transfer of R-factors from transient to persistent enterobacterial strains with good colonization capabilities, and (6) recombination between R-factors and plasmids and/or transposons mediating colonization capability or virulence.

A. Animal to Man Transmission of R-Factors

The incidence of antibiotic resistance in the common meat animals is generally much higher than that found in nonhospitalized persons. Detailed studies by Linton[152] and Jackson[55] in the U.K. have established beyond doubt that high levels of resistant *E. coli* occur in pigs, poultry, and calves, with insignificant numbers occurring in adult cattle and sheep. Linton,[50] in a recent review, states that it is now the norm to find the larger proportion of *E. coli* in pigs, poultry, and calves resistant to one or more of the therapeutically useful antibiotics, whether or not the animals are receiving antibiotics. Using the shared characters of antibiotic resistance and O-serotype, Linton and co-workers[152] clearly demonstrated that resistant *E. coli* from the gut were distributed on the animal carcasses during slaughter and subsequent processing and packaging of the meat. Studies by a variety of workers on poultry meat utilized in hospitals[153] and purchased from retail shops for domestic use[154] confirmed that *E. coli* of animal origin without doubt reach the consumer, especially via the kitchen.

The colonization of the human gut by *E. coli* strains of animal origin is more difficult to prove. Linton and co-workers[152] studied the serotype distribution of *E. coli* in man and animals and concluded that, based on O-serotyping alone, it was not possible to distinguish normal gut *E. coli* of animals from those commonly present in man.[152] In a study of thousands of isolates, the range of antibiotic-resistant *E. coli* O-serotypes from pigs, calves, and poultry was found to be closely similar to the O-serotype range representative of domiciliary humans.[152] Evidence for the persistence of animal strains of *E. coli* in the human gut was obtained by Cooke and co-workers[155] following ingestion of a very large dose of a marked

E. coli strain. Linton and co-workers[156] studied the colonization of human volunteers by *E. coli* from commercial chicken carcases. Clear evidence was obtained that one of the volunteers, in the absence of any antibiotic selection, became colonized by resistant *E. coli* of chicken origin (five different O-serotypes of chicken origin were detected) to a point at which the strains became a major part of the fecal flora for a period of several days.[152,156] Interestingly, colonization was shown to have occurred on the day after the carcase was handled, prepared, and cooked but before it was eaten, indicating that it was the handling of the raw meat that provided the opportunity for onward transmission of the animal *E. coli* strains.

It is clear, therefore, from these and other related studies that *E. coli* of animal origin are transmitted to man via the food chain. The similarity between the O-serotypes of *E. coli* in man and animals and the incontrovertible evidence provided by Linton's seminal paper[156] strongly suggest that R$^+$ *E. coli* of animal origin can colonize the human intestine and may persist within the gut for significant periods of time, even in the absence of any selective pressure by antibiotic administration.

B. In Vivo Transfer of R-Factors

There is no evidence to suggest that the acquisition by man of resistant *E. coli* of animal origin or their persistence for long periods in the gut produces any detectable pathological effects on the human host.[152] Of far greater concern, therefore, is the potential transfer of R-factors from transient R$^+$ animal *E. coli*, or from more persistent animal serotypes which temporarily colonize the gut, to the endogenous *E. coli* strains or to pathogenic species such as *Salmonella* or *Shigella*. Normally, in healthy humans, one or a few *E. coli* serotypes predominate in the gut at one time. Some serotypes persist for many months whereas others occur more transiently.[157] The ability of a particular strain to colonize the intestine depends in part on the details of its surface structure.[158]

Obviously, the transfer of R-factors to effective colonizers of the human gut would ensure longer-term residence of the R-factor within the gut.

The extent to which R-factor transfer occurs within the gut has been the focus of considerable controversy. Broda[158] states that the conditions in the normal gut (alkaline pH, bile salts, fatty acids, anaerobiosis, etc.) are not conducive towards conjugation. Also, enterobacteria constitute less then 1% of the total flora in the gut so that their numbers are considerably lower than the numbers of *E. coli* used in in vitro conjugation experiments in the laboratory.[158] Although conjugation was considered to be inefficient under anaerobic conditions, more recent studies show that, for some plasmids, it proceeds as efficiently as under aerobic conditions.[159] A recent study by Singleton and Anson[160] concluded that conjugal transfer could take place over a relatively wide pH range, with the optimum pH varying with the plasmid under study and with the characteristics of the donor and recipient cells.

Anderson and co-workers demonstrated transfer between *E. coli* strains in the human intestine. However, this could be detected only in the presence of an antibiotic which allowed an increase in the number of R$^+$ cells.[161] Molecular studies confirmed that the R-factors acquired by the recipient cells were indeed those present originally only in the donor strain.[161] Petrocheilou et al.[162] succeeded in demonstrating transfer between *E. coli* strains in the gut of a healthy volunteer in the absence of antibiotic selection. Since 202 days elapsed before the R-factor was first detected in another strain, Petrocheilou concluded that R-factor transfer may not occur commonly in the gut. On the other hand, R-factor transfer between *E. coli* strains in the rumen of sheep occurs efficiently after 24 hr starvation,[163] in mice otherwise free of bacteria,[164] in gnotoxenic chickens,[145] and in the digestive tracts of rats, pigs, cows, and turkeys.[166]

Clearly, in vivo transfer of R-factors between *E. coli* strains can occur within the intestine of both man and animals. To what extent it occurs and to what extent it contributes to the current incidence of R-factors in enterobacterial species is unclear and may, in fact, be virtually impossible to ascertain.

C. R-Factor Transfer Under Environmental Conditions

Studies on the survival of R-factor-carrying coliforms in sewage, sewage treatment plants, and in fecally contaminated waters and sediments have indicated no significant difference between the survival of R^+ and R^- bacteria.[167,168] In conventional sewage treatment plants, quiescent processes such as sedimentation were observed by Grabow and co-workers[169] to favor conjugation and transfer of resistance between enterobacterial strains. Mach and Grimes[170] confirmed that wild-type R^+ strains of *S. enteritidis, Proteus mirabilis,* and *E. coli* could transfer their R-factors to *E. coli* and *Shigella sonnei* in membrane diffusion chambers which were placed in the primary and secondary settling tanks of a wastewater treatment plant. R-factor transfer has also been shown to occur in dialysis bags immersed in a river and dam under environmental conditions.[171] Cooke studied the transferability of a random collection of R^+ *E. coli* strains (isolated from environmental sources) at a variety of temperatures, ranging from 15 to 35°C, in nutrient broth, river water, and seawater and concluded that almost 50% of the isolates could transfer their R-factors under environmental nutrient and temperature conditions.[7]

It is clear from these studies that transfer of R-factors can occur under environmental conditions. However, the frequency of transfer reported was low and the experimental conditions utilized generally involved relatively high numbers of donor and recipient cells ($>10^6$ per milliliter) which were placed in water-permeable containment devices in the river, treatment plant, or other environments under study. Low cell counts, water turbulence, the presence of antibacterial compounds, and a variety of other chemical and physical factors may significantly affect the rate of R-factor transfer under natural conditions between wild-type strains. Nevertheless, environments, such as the quiescent areas of settling tanks, may present sufficiently high numbers of bacteria in close enough proximity to effect significant transfer of R-factors at ambient temperatures.

The above discussion clearly shows that R-factors which emerge in the microbial flora of animals as a consequence of agricultural and veterinary usage of antibiotics may reach the human population via the food chain and either colonize the human gut or transfer in vivo to indigenous commensal strains. Opportunities for further transfer are provided in voided faeces and in sewage treatment plants. In addition, the onward transmission of R^+ strains from person to person has been clearly demonstrated by the studies of Petrocheilou and co-workers.[172] Broda[158] concludes that it may now be impossible to avoid being colonized by R^+ bacteria. Obviously, any factor which increases this flow of R-factors from animals to man further compromises the therapeutic use of antibiotics in human medicine.

XVII. ENVIRONMENTAL ASPECTS OF H PLASMID TRANSFER

Since thermosensitive R-factors of the Inc H group transfer optimally at temperatures below 30°C, the possibility exists that conjugal transfer under environmental conditions may be instrumental in the spread of H plasmids between bacterial species and ultimately between their human and animal hosts.

Very few studies have been carried out to date on H plasmid transfer under environmental conditions. Linton and co-workers[15] noted that calves, infected with a HI2 plasmid-carrying, chloramphenicol-resistant strain of *S. typhimurium* in a calf unit in Somerset, U.K., invariably harbored *E. coli* resistant to the same range of antibiotics. The pattern of excretion of resistant coliforms and the combination of R-determinants carried by 89% of the chloramphenicol-resistant *E. coli* suggested that the HI2 plasmid had been transferred from *Salmonella* to indigenous *E. coli* strains. Since the HI2 plasmid was detected in six different

E. coli serotypes, it was apparent that multiple transfers had occurred — either between *Salmonella* and *E. coli* or between H⁺ *E. coli* transconjugants and other sensitive *E. coli* strains. Despite repeated attempts, Linton and co-workers could not demonstrate H plasmid transfer in muzzled calves which had no access to their feces. By contrast, transfer was readily detected in unmuzzled calves dosed with genetically marked strains containing the HI2 plasmid.[151] Transfer was also readily demonstrated in voided feces held at 28°C. Linton[15,151] concluded that transfer of the H plasmids had occurred outside the animal in the fecally contaminated environment and colonization of the animals occurred by the feco-oral route.

In an ongoing study of H plasmid transfer under environmental conditions, Maher and Colleran demonstrated transfer of HI2 plasmids from *E. coli* to marked sensitive strains of *E. coli* and *S. typhimurium* in sewage and raw milk held at 22 and 26°C.[174] Transfer was detected in 24-hr mating experiments with cellular levels of donor and recipient of 1×10^6 per milliliter. Conjugation did not occur at donor and recipient cell numbers below 10^6 per milliliter, even when the mating period was extended to several days. Further studies are underway to determine H plasmid transfer frequencies in settling tanks of wastewater treatment plants and in a variety of other environments at ambient temperatures.

Epidemiological studies have highlighted the predominance of Inc H plasmids in the salmonellae. Although the data available are limited, clear evidence has also been presented on their emerging prevalence in other enteric bacteria in certain geographical regions. The variety of genes now known to be encoded by H plasmids, i.e., antibiotic and metal multiresistance, substrate utilization, colonization/persistence capability, etc., coupled with their unique thermosensitive conjugal transfer characteristics emphasizes the urgent need for detailed biochemical and genetic characterization of this plasmid group. Recent advances in molecular biology and genetic engineering have provided a battery of techniques whereby these unusually large plasmids may be fully investigated.

Of particular interest with respect to the ecology of H plasmids is the observed rapid dissemination of similar H plasmids in *Salmonella* isolates from different continents.[39,41-43] Investigation of the biochemical basis of the resistance mechanisms determined by these plasmids, coupled with genetic studies of the gene loci on the plasmid DNA, is required in order to establish, with certainty, the relatedness of H plasmids from different geographical regions. Transposability studies and determination of the extent of homology between Inc H transposons and those of other plasmid groups is also required in order to establish the relationship between Inc H and other plasmid types and the origin of the resistance determinants presently carried by plasmids of the H group.

Molecular and genetic techniques may also be utilized to investigate the unusual thermosensitive transfer system of Inc H plasmids. Taylor and co-workers' pioneering studies in this field have already identified at least two separate regions involved in H plasmid transfer. Further studies, involving the cloning of H plasmid DNA fragments, hybridization with labeled probes, and biochemical analysis of tra gene products, are necessary in order to determine the basis for the thermosensitivity of H plasmid transfer and delineate, in detail, the underlying biochemical steps involved in the transfer process.

Recent studies have highlighted the possibility that Inc H plasmids may carry colonization genes which enhance the persistence of the host cell in the intestine.[151] Because of the veterinary and medical significance of this trait, molecular, genetic and biochemical analysis of this determinant is urgently required. The large size of H plasmid DNA also leads to speculation that other, as yet unknown, traits may be encoded by plasmids of the Inc H group. These characteristics, if present, must be identified and the determinants analysed by molecular and genetic techniques.

The thermosensitive transfer system encoded by Inc H plasmids is their single most important characteristic. Ecological studies must, in future, therefore, include assessment of the significance of thermosensitive transfer in a variety of natural environments. Such studies may identify a need for re-evaluation of the standards for wastewater treatment systems and for effluent discharge into natural waters. As stated by Linton and co-workers[15] in a recent paper, should the indigenous *E. coli* of animals and man become colonized generally by H plasmids, as a consequence of thermosensitive transfer under environmental conditions and because of antibiotic selection pressure and other factors, the potential for serious animal and public health problems cannot be too strongly emphasized.

REFERENCES

1. **Watanabe, T.,** Infective heredity of multiple drug resistance in bacteria, *Bacteriol. Rev.,* 27, 87, 1963.
2. **Falkow, S.,** *Infectious Multiple Drug Resistance,* Pion, London, 1975, chap. 5.
3. **Richmond, M. H.,** Some environmental consequences of the use of antibiotics, *J. Appl. Bact.,* 35, 155, 1972.
4. **Linton, A. H.,** Antibiotics, animals and man — an appraisal of a contentious subject, in *Antibiotics and Antibiosis in Agriculture,* Woodbine, M., Ed., Butterworths, London, 1977, 315.
5. **Lacey, R. W.,** A critical appraisal of the importance of R-factors in the Enterobacteriaceae *in vivo, J. Antimicrob. Chemother.,* 1, 25, 1975.
6. **Gardner, P. and Smith, D. H.,** Studies on the epidemiology of resistance (R) factors, *Ann. Intern. Med.,* 71, 1, 1969.
7. **Cooke, M. D.,** R-factor transfer under environmental conditions, in *Proc. 4th Int. Symp. on Antibiotic Resistance: Transposition and Other Mechanisms,* Mitsuhashi, S., Rosival, L., and Kremery, V., Eds., Springer-Verlag, 1980, 203.
8. **Anderson, E. S.,** The problem and implications of chloramphenicol resistance in the typhoid bacillus, *J. Hyg. Camb.,* 74, 289, 1975.
9. **Grindley, N. D. F., Grindley, J. N., and Anderson, E. S.,** R-factor compatibility groups, *Mol. Gen. Genet.,* 119, 287, 1972.
10. **Anderson, E. S. and Smith, H. R.,** Chloramphenicol resistance in the typhoid bacillus, *Br. Med. J.,* 3, 329, 1972.
11. **Threlfall, E. J.,** Epidemic spread of a chloramphenicol resistant strain of *S. typhimurium* phage type 204 in bovine animals in Britain, *Vet. Record,* 103, 438, 1978.
12. **Threlfall, E. J., Ward, L. R., and Rowe, B.,** Spread of multiresistant strains of *S. typhimurium* phage types 204 and 193 in Britain, *Br. Med. J.,* ii, 997, 1978.
13. **Threlfall, E. J.,** Multiresistant epidemic strains of *S. typhimurium* in Britain, in *Resistance and Pathogenic Plasmids, Proc. of CEC Seminar, Brussels, 15—16 October, 1981,* Pohl, P. and Leunen, J., Eds., National Institute for Veterinary Research, Brussels, 1982, 103.
14. **Threlfall, E. J., Ward, L. R., Ashley, A. S., and Rowe, B.,** Plasmid encoded trimethoprim resistance in multiresistant epidemic *S. typhimurium* phage types 204 and 193 in Britain, *Br. Med. J.,* i, 1210, 1980.
15. **Linton, A. H., Timoney, J. F., and Hinton, M.,** The ecology of chloramphenicol-resistance in *S. typhimurium* and *E. coli* in calves with endemic *Salmonella* infection, *J. Appl. Bact.,* 50, 115, 1981.
16. **Datta, N.,** Epidemiology and classification of plasmids, in *Microbiology—1974,* Schlessinger, D., Ed., American Society for Microbiology, Washington, D.C., 1975, 9.
17. **Anderson, E. S. and Threlfall, E. J.,** The characterisation of plasmids in the enterobacteria, *J. Hyg. Camb.,* 72, 471, 1974.
18. **Grindley, N. D. F., Humphreys, G. O., and Anderson, E. S.,** Molecular studies of R-factor compatibility groups, *J. Bacteriol.,* 115, 387, 1973.
19. **Smith, H. R., Grindley, N. D. F., Humphreys, G. O., and Anderson, E. S.,** Interactions of group H resistance factors with the F-factor, *J. Bacteriol.,* 115, 623, 1973.
20. **Taylor, D. E. and Grant, R. B.,** Incompatibility and surface exclusion properties of H_1 and H_2 plasmids, *J. Bacteriol.,* 131, 174, 1977.
21. **Bradley, D. E., Hughes, V. M., Richards, H., and Datta, N.,** R-plasmids of a new incompatibility group determine constitutive production of H pili, *Plasmid,* 7, 230, 1982.
22. **Hedges, R. W., Rodriguez-Lemoine, V., and Datta, N.,** R-factors from *Serratia marcescens, J. Gen. Microbiol.,* 86, 88, 1975.
23. **Minor, L. Le, Coynault, C., Chabbert, Y., Gerbaud, G., and Minor, S. Le,** Groupes de compatibilite de plasmides metaboliques, *Ann. Microbiol. (Inst. Pasteur),* 127B, 31, 1976.
24. **Roussel, A. F. and Chabbert, Y. A.,** Taxonomy and epidemiology of Gram-negative bacterial plasmids studies by DNA-DNA filter hybridization in formamide, *J. Gen. Microbiol.,* 104, 269, 1978.

25. **Taylor, D. E. and Levine, J. G.,** Studies of temperature-sensitive transfer and maintenance of H incompatibility group plasmids, *J. Gen. Microbiol.,* 116, 475, 1980.

26. **Bradley, D. E., Taylor, D. E., and Cohen, D. R.,** Specification of surface mating systems among conjugative drug resistance plasmids in *E. coli* K-12, *J. Bacteriol.,* 143, 1466, 1980.

27. **Sasakawa, C., Takamatsu, N., Danbara, H., and Yoshikawa, M.,** A method of plasmid classification by integrative incompatibility, *Plasmid,* 3, 116, 1980.

28. **Yoshida, T., Takahashi, I., Tubahara, H., Sasakawa, C., and Yoshikawa, M.,** Significance of filter mating in integrative incompatibility tests for plasmid classification, *Microbiol. Immunol.,* 28, 63, 1984.

29. **Palomares, J. C. and Perea, E. J.,** A rapid method for determining the incompatibility group of R plasmids, *Plasmid,* 4, 352, 1980.

30. **Nakaya, R., Nakamura, A., and Murata, T.,** Resistance transfer agents in *Shigella, Biochem. Biophys. Res. Commun.,* 3, 654, 1960.

31. **Watanabe, T., Nishida, H., Ogata, C., Arai, T., and Sato, S.,** Episome-mediated transfer of drug resistance in *Enterobacteriaceae.* VII. Two types of naturally occurring R-factor, *J. Bacteriol.,* 88, 716, 1964.

32. **Lawn, A. M., Meynell, E., Meynell, G. G., and Datta, N.,** Sex pili and the classification of sex factors in *Enterobacteriaceae, Nature (London),* 216, 343, 1967.

33. **Meynell, E.,** Pseudo-fi⁺ I-like sex factor, RG2 (1), selective for increased pilus synthesis, *J. Bacteriol.,* 113, 302, 1973.

34. **Datta, N. and Olarte, J.,** R factors in strains of *Salmonella typhi* and *Shigella dysenteriae* I isolated during epidemics in Mexico: classification by compatibility, *Antimicrob. Agents Chemother.,* 5, 310, 1974.

35. **Smith, H., Williams, and Parsell, Z. E.,** Transmissible substrate-utilizing ability in enterobacteria, *J. Gen. Microbiol.,* 87, 129, 1975.

36. **Pohl, P., Thomas, J., Moury, J., Van Robaeys, G., and Dreze, P.,** Plasmides codant pour la resistance au chloramphenicol chez des Salmonella epidemiques, *Ann. Med. Vet.,* 124, 47, 1980.

37. **Makino, S., Ishiguro, N., and Sato, G.,** Change of drug resistance patterns and genetic properties of R plasmids in *Salmonella typhimurium* of bovine origin isolated from 1970 to 1979 in Northern Japan, *J. Hyg. Camb.,* 87, 257, 1981.

38. **Bezanson, G. S., Pauze, M., and Lior, H.,** Antibiotic resistance and R-plasmids in food-chain *Salmonella:* evidence of plasmid relatedness, *Appl. Environ. Microbiol.,* 41, 585, 1981.

39. **Anderson, E. S.,** The geographical predominance of resistance transfer systems of various compatibility groups in Salmonellae, in *Topics Infectious Disease,* Vol. 2, Drews, J. and Hogenauer, G., Eds., Springer-Verlag, Berlin, 1977, 25.

40. **Avril, J. L., Dabernat, H. J., Gerbaud, G. R., Horodniceanu, T., Lambert-Zechovsky, N., Le Minor, S., Mendez, B., and Chabbert, Y. A.,** Groupes d'incompatibilite des plasmides R chez les souches de *Salmonella* epidemiques, *Ann. Microbiol. (Inst. Pasteur),* 128B, 12, 1977.

41. **Pohl, P., Lintermans, P., Corbion, B., Gledel, J., Le Minor, L., Chasseur, M.-L., and Ghysels, G.,** Plasmides du groupe d'incompatibilite H chez des souches de *Salmonella* multiresistantes, *Ann. Microbiol. (Inst. Pasteur),* 132B, 399, 1981.

42. **Grant, R. B. and di Mambro, L.,** Antimicrobial resistance and resistance plasmids in *Salmonella* from Ontario, Canada, *Can. J. Microbiol.,* 23, 1266, 1977.

43. **Taylor, D. E. and Grant, R. B.,** Inhibition of bacteriophage lambda, T1 and T7 development by R plasmids of the H incompatibility group, *Antimicrob. Agents Chemother.,* 10, 762, 1976.

44. **Taylor, D. E., Shermer, M., and Grant, R. B.,** Incidence of the HI2 group of plasmids in chloramphenicol-sensitive *Salmonella* isolated in 1974 from clinical sources in Ontario, *Can. J. Microbiol.,* 24, 600, 1978.

45. **Taylor, D. E., Levine, J. G., and Kouvelos, K. L.,** Incidence of plasmid DNA in *Salmonella* strains isolated from clinical sources in Ontario, Canada, during 1979 and 1980, *Can. J. Microbiol.,* 28, 1150, 1982.

46. **Terakado, N., Ohya, T., Ueda, H., and Isayama, Y.,** A survey on drug resistance and R-plasmids in *Salmonella* isolated from domestic animals in Japan, *Jpn. J. Vet. Sci.,* 42, 543, 1980.

47. **Ishiguro, N., Makino, S., Sato, G., and Hashimoto, K.,** Antibiotic resistance and genetic properties of R-plasmids in *Salmonella* isolates of swine origin in Japan, *Am. J. Vet. Res.,* 41, 46, 1980.

48. **Yoshida, T., Terawaki, Y., and Nakaya, R.,** R-plasmids with thermosensitive transferability in *Salmonella* strains isolated from humans, *Microbiol. Immunol.,* 22, 735, 1978.

49. **Niida, M., Makino, S., Ishiguro, N., Sato, G., and Nishio, T.,** Genetic properties of conjugative R-plasmids in *E. coli* and *Salmonella* isolates from feral and domestic pigeons, crows, and kites, *Zbl. Bakt. Hyg. I. Abt. Orig.,* A255, 271, 1983.

50. **Linton, A. H.,** The Swann Report and its impact, in *The Control of Antibiotic Resistant Bacteria,* Stuart-Harris, C. and Harris, D. M., Eds., Academic Press, London, 1982, 183.

51. **Ishiguro, N., Goto, J., and Sato, G.,** Genetic relationship between R-plasmids derived from *Salmonella* and *E. coli* obtained from a pig farm, and its epidemiological significance, *J. Hyg. Camb.,* 84, 365, 1980.

52. **Timoney, J. F.,** The epidemiology and genetics of antibiotic resistance of *Salmonella typhimurium* isolated from diseased animals in New York, *J. Infect. Dis.,* 137, 67, 1978.

53. **Timoney, J. F.,** R plasmids in pathogenic enterobacteriaceae from calves, in *Molecular Biology, Pathogenicity and Ecology of Bacterial Plasmids,* Levy, S. B., Clowes, R. C., and Koenig, E. L., Eds., Plenum Press, London, 1981, 547.

54. **Voogd, C. E., van Leeuwen, W. J., Guinee, P. A. M., Manten, A., and Valkenburg, J. J.,** Incidence of resistance to ampicillin, chloramphenicol, kanamycin and tetracycline among *Salmonella* species isolated in the Netherlands in 1972, 1973 and 1974, *Antonie van Leeuwenhoek J. Microbiol. Serol.,* 43, 269, 1977.

55. **Jackson, G.,** A survey of antibiotic resistance of *Escherichia coli* isolated from farm animals in Great Britain from 1971 to 1977, *Vet. Record,* 108, 325, 1981.

56. **Wilcock, B. and Olander, H.,** Influence of oral antibiotic feeding on the duration and severity of clinical disease, growth performance and pattern of shedding in swine inoculated with *Salmonella typhimurium, J. Am. Vet. Med. Assoc.,* 172, 472, 1978.

57. **Smith, H. Williams, Parsell, Z., and Green, P.,** Thermosensitive antibiotic resistance plasmids in enterobacteria, *J. Gen. Microbiol.,* 109, 37, 1978.

58. **Rangnekar, V. M., Banker, D. D., and Jhala, H. I.,** Drug resistance and incompatibility groups of R-plasmids in intestinal *Escherichia coli, Ind. J. Med. Res.,* 75, 492, 1982.

59. **Terakado, N. and Sato, G.,** Demonstration of the so-called Mexican type R plasmids in *Escherichia coli* isolates from domestic animals and pigeons, *Microbiol. Immunol.,* 22, 227, 1978.

60. **Nakamura, M., Fukazawa, M., Hoshimura, H., and Koeda, T.,** Genetic properties of R plasmids derived from *Escherichia coli* strains isolated from imported mynahs, *Jpn. J. Vet. Sci.,* 45, 39, 1983.

61. **Romero, E., Pagani, L., and Perduca, M.,** Compatibility grouping of chloramphenicol-resistant fi⁻ R-factors isolated in Italy, *Microbiologica,* 1, 51, 1978.

62. **Maher, D. and Colleran, E.,** Thermosensitive plasmids in coliform bacteria, *Ir. J. Med. Sci.,* 153, 40, 1984.

63. **Taylor, D. E. and Grant, R. B.,** Incompatibility and bacteriophage inhibition properties of N-I, a plasmid belonging to the H₂ incompatibility group, *Mol. Gen. Genet.,* 153, 5, 1977.

64. **Frost, J. A. and Rowe, B.,** Plasmid-mediated antibiotic resistance in *Shigella flexneri* isolated in England and Wales between 1974 and 1978, *J. Hyg. Camb.,* 90, 27, 1983.

65. **Taylor, D. E. and Grant, R. B.,** R plasmids of the S incompatibility group, *Antimicrob. Agents Chemother.,* 12, 431, 1977.

66. **Taylor, D. E. and Summers, A. O.,** Association of tellurium resistance and bacteriophage inhibition conferred by R plasmids, *J. Bacteriol.,* 137, 1430, 1979.

67. **Richards, H. and Datta, N.,** Plasmids and transposons acquired by *Salmonella typhi* in man, *Plasmid,* 8, 9, 1982.

68. **Smith, H. Williams,** Thermosensitive transfer factors in chloramphenicol-resistant strains of *Salmonella typhi,* Lancet, ii, 281, 1974.

69. Surveillance for the Prevention and Control of Health Hazards due to Antibiotic-Resistant Enterobacteria, *WHO Tech. Rep. Ser.,* 624, 21, 1978.

70. **Frost, J. A., Threlfall, E. J., and Willshaw, G. A.,** Methods of studying transferable resistance to antibiotics, in *Antibiotics: Assessment of Antimicrobial Activity and Resistance,* Russell, A. and Quesnel, L. B., Eds., Academic Press, London, 1983, 265.

71. **Rodriguez-Lemoine, V., Jacob, A. E., Hedges, R. W., and Datta, N.,** Thermosensitive production of their transfer systems by group S plasmids, *J. Gen. Microbiol.,* 86, 111, 1975.

72. **Taylor, D. E. and Levine, J. G.,** Studies of temperature-sensitive transfer and maintenance of H incompatibility group plasmids, *J. Gen. Microbiol.,* 116, 475, 1980.

73. **Yoshida, H. and Nakatani, R.,** R-plasmid whose conjugation transfer process is temperature-sensitive (in Japanese), *Jpn. J. Bacteriol.,* 31, 192, 1976.

74. **Bradley, D. E., Taylor, D. E., and Cohen, D. R.,** Specification of surface mating systems among conjugative drug resistance plasmids in *Escherichia coli* K-12, *J. Bacteriol.,* 143, 1466, 1980.

75. **Dennison, S. and Baumberg, S.,** Conjugational behaviour of N plasmids in *Escherichia coli, Mol. Gen. Genet.,* 138, 323, 1975.

76. **Bradley, D. E. and Choudhari, T.,** Characteristics and interaction with bacteriophages of pili determined by a plasmid of the N incompatibility group, in *Plasmids and Transposons,* Stuttard, C. and Rozee, K. R., Eds., Academic Press, London, 1980, 335.

77. **Grant, R. B., Bannatyne, R. M., and Shapley, A. J.,** Chloramphenicol and ampicillin-resistant *Salmonella typhimurium* in Ontario, *J. Infect. Dis.,* 134, 354, 1976.

78. **Yoshida, T., Takahashi, I., Tubahara, H., Sasakawa, C., and Yoshikawa, M.,** Significance of filter mating in integrative incompatibility test for plasmid classification, *Microbiol. Immunol.,* 28, 63, 1984.

79. **Rodriguez-Lemoine, V. and Cavazza, M. E.,** On the transfer system determined by plasmids belonging to incompatibility group S, in *Molecular Biology, Pathogenicity and Ecology of Bacterial Plasmids,* Levy, S. B., Clowes, R. C. and Koenig, E. L., Eds., Plenum Press, New York, 1981, 649.

80. **Broda, P.,** Conjugation in bacteria, in *Plasmids of Medical, Environmental and Commercial Importance,* Timmis, K. N. and Puhler, A., Eds., Elsevier/North-Holland, New York, 1979, 47.

81. **Willetts, N. and Shurray, R.,** The conjugation system of F-like plasmids, *Ann. Rev. Genet.,* 14, 41, 1980.

82. **Willetts, N. and Wilkins, B.,** Processing of plasmid DNA during bacterial conjugation, *Microbiol. Rev.,* 48, 24, 1984.

83. **Thatte, V. and Iyer, V. N.,** Cloning of a plasmid region specifying the N transfer system of bacterial conjugation in *Escherichia coli, Gene,* 21, 227, 1983.

84. **Moore, R. J. and Krishnapilla, V.,** Tn7 and Tn501 insertions into *Pseudomonas aeruginosa* plasmid R91-5: mapping of two transfer regions, *J. Bacteriol.,* 149, 276, 1982.

85. **Barth, P. T., Grinter, N. J., and Bradley, D. E.,** Conjugal transfer system of plasmid RP4: analysis by transposon 7 insertion, *J. Bacteriol.,* 133, 43, 1978.

86. **Barth, P. T.,** RP4 and R300B as wide host range plasmid cloning vehicles, in *Plasmids of Medical, Environmental and Commercial Importance,* Timmis, K. N. and Puhler, A., Eds., Elsevier/North-Holland, New York, 1979, 399.

87. **Holsters, M., Silva, B., Van Vliet, F., Genetello, C., de Block, M., Dhaese, P., Depicher, A., Inze, D., Engler, G., Villarroel, R., Van Montague, M., and Schell, J.,** The functional organization of the nopaline *A. tumefaciens* plasmid, pTiC58, *Plasmid,* 3, 212, 1980.

88. **Taylor, D. E.,** Transfer-defective and tetracycline-sensitive mutants of the incompatibility group H1 plasmid R27 generated by insertion of transposon 7, *Plasmid,* 9, 227, 1983.

89. **Whitely, M. and Taylor, D. E.,** Identification of DNA homologies among H incompatibility group plasmids by restriction enzyme digestion and southern transfer hybridization, *Antimicrob. Agents Chemother.,* 24, 194, 1983.

90. **Jacob, A. E., Shapiro, J. A., Yamamoto, L., Smith, D. I., Cohen, S. N., and Berg, D.,** Plasmids studied in *Escherichia coli* and other enteric bacteria, in *DNA Insertion Elements, Plasmids and Episomes,* Bukhari, A. I., Shapiro, J. A., and Adhya, S. L., Eds., Cold Spring Harbor Laboratory, Cold Spring Harbor, N.Y., 1977, 607.

91. **Datta, N.,** Plasmids of enteric bacteria, in *Antimicrobial Drug Resistance,* Bryan, L. E., Ed., Academic Press, London, 1984, 487.

92. **Taylor, D. E. and Grant, R. B.,** Bacteriophage inhibition properties of H2 plasmids, in *Microbiology — 1978,* Schlessinger, D., Ed., American Society for Microbiology, Washington, D.C., 1978, 214.

93. **Humphreys, G. O., Willshaw, G. A., and Anderson, E. S.,** A simple method for the preparation of large quantities of pure plasmid DNA, *Biochim. Biophys. Acta,* 383, 457, 1975.

94. **Hansen, J. B. and Olsen, R. H.,** Isolation of large bacterial plasmids and characterization of the P2 incompatibility group plasmids pMG1 and pMG5, *J. Bacteriol.,* 135, 227, 1978.

95. **Currier, T. C. and Nester, E. W.,** Isolation of covalently closed circular DNA of high molecular weight from bacteria, *Anal. Biochem.,* 76, 431, 1976.

96. **Hirsch, P. R., Van Montagu, M., Johnston, A. W. B., Brewin, N. J., and Schell, J.,** Physical identification of bacteriocinogenic, nodulation and other plasmids in strains of *Rhizobium leguminosarum, J. Gen. Microbiol.,* 120, 403, 1980.

97. **Prakash, R. K., Schilperoort, R. A., and Nuti, M. P.,** Large plasmids of fast-growing rhizobia: homology studies and location of structural nitrogen fixation (nif) genes, *J. Bacteriol.,* 145, 1129, 1981.

98. **Sansonetti, P. J., d'Hauteville, H., Ecobichon, C., and Pourcel, C.,** Molecular comparison of virulence plasmids in *Shigella* and enteroinvasive *Escherichia coli, Ann. Microbiol. (Inst. Pasteur),* 134A, 295, 1983.

99. **Grindley, N. D. F., Humphreys, G. O., and Anderson, E. S.,** Molecular studies of R-factor compatibility groups, *J. Bacteriol.,* 115, 387, 1973.

100. **Anderson, E. S., Humphreys, G. O., and Willshaw, G. A.,** The molecular relatedness of R-factors in Enterobacteria of human and animal origin, *J. Gen. Microbiol.,* 91, 387, 1975.

101. **Bradley, D. E.,** Morphological and serological relationships of conjugative pili, *Plasmid,* 4, 155, 1980.

102. **Bradley, D. E.,** Determination of pili by conjugative bacterial drug resistance plasmids of incompatibility groups B, C, H, J, K, M, V and X, *J. Bacteriol.,* 141, 828, 1980.

103. **Watanabe, T., Nishida, H., Ogata, C., Arai, T., and Sato, S.,** Episome-mediated transfer of drug resistance in Enterobacteriaceae. VII. Two types of naturally-occurring R factors, *J. Bacteriol.,* 88, 716, 1964.

104. **Watanabe, T., Takano, T., Arai, T., Nishida, H., and Sato, S.,** Episome-mediated transfer of drug resistance in Enterobacteriaceae, *J. Bacteriol.,* 92, 477, 1966.

105. **Rouland-Dussoix, D., Yoshimari, R., Greene, P., Betlach, M., Goodman, H., and Boyer, H.,** R factor controlled restriction and modification of deoxyribonucleic acid, in *Microbiology — 1974,* Schlessinger, D., Ed., American Society of Microbiology, Washington D.C., 1975, 187.

106. **Humphreys, G. O.,** The molecular nature of R-factors in different bacterial hosts, in *Topics Infectious Disease,* Vol. 2, Drews, J. and Hogenauer, G., Eds., Springer-Verlag, Berlin, 1977, 277.

107. **Rangnekar, V. H., Gadre, S. V., Mukerji, S., and Chitnis, D. S.,** Genetic stability of Inc H1 and other incompatibility group R-plasmids in *Salmonella typhi, Ind. J. Med. Res.,* 76, 512, 1982.

108. **Timoney, J. F., Taylor, D. E., Shin, S., and McDonagh, P.,** pJT2: unusual H1 plasmid in a highly virulent lactose-positive and chloramphenicol-resistant *Salmonella typhimurium* strain from calves, *Antimicrob. Agents Chemother.,* 18, 480, 1980.

109. **Molina, A. M., Musmanno, R. A., Margolicci, M. A., and Andreoni, O.,** An H2 plasmid determining lactose fermentation and drug resistance in a strain of *Salmonella infantis* and its recombination with an FII plasmids, *Microbiologica,* 4, 281, 1981.

110. **Johnson, E. M., Wohlhteier, J. A., Placek, B. P., Sleet, R. B., and Baron, L. S.,** Plasmid-determined ability of a *Salmonella tennessee* strain to ferment lactose and sucrose, *J. Bacteriol.,* 125, 385, 1976.

111. **Hall, B. G.,** Chromosomal mutation for citrate utilization by *Escherichia coli* K-12, *J. Bacteriol.,* 151, 269, 1982.

112. **Sato, G., Asagi, M., Oka, C., Ishiguro, N., and Terakado, N.,** Transmissible citrate-utilizing ability in *Escherichia coli* isolated from pigeons, pigs and cattle, *Microbiol. Immunol.,* 22, 357, 1978.

113. **Ishiguro, N., Oka, C., and Sato, G.,** Isolation of citrate-positive variants of *Escherichia coli* from domestic pigeons, cattle and horses, *Appl. Environ. Microbiol.,* 36, 217, 1978.

114. **Smith Williams, H., Parsell, Z., and Green, P.,** Thermosensitive plasmids determining citrate utilization, *J. Gen. Microbiol.,* 109, 305, 1978.

115. **Ishiguro, N., Oka, C., Hanzawa, Y., and Sato, G.,** Plasmids in *Escherichia coli* controlling citrate-utilizing ability, *Appl. Environ. Microbiol.,* 38, 956, 1979.

116. **Ishiguro, N., Hirose, K., Asagi, M., and Sato, G.,** Incompatibility of citrate-utilization plasmids isolated from *Escherichia coli, J. Gen. Microbiol.,* 123, 193, 1981.

117. **Ishiguro, N., Hirose, K., and Sato, G.,** Distribution of citrate-utilization plasmids in *Salmonella* strains of bovine origin in Japan, *Appl. Environ. Microbiol.,* 40, 446, 1980.

118. **Ishiguro, N., Oka, C., Hanzawa, Y., and Sato, G.,** Isolation of a citrate-utilization plasmid from a bovine *Salmonella typhimurium* strain, *Microbiol. Immunol.,* 24, 757, 1980.

119. **Ishiguro, N., Sato, G., Ichizo, S., and Konishi, T.,** Incompatibility of R plasmids derived from *Salmonella* and *Escherichia coli* strains isolated simultaneously from a bovine faecal sample, *Am. J. Vet. Res.,* 41(12), 1982, 1980.

120. **Ishiguro, N. and Sato, G.,** Properties of a transmissible plasmid conferring citrate-utilizing ability in *Escherichia coli* of human origin, *J. Gen. Microbiol.,* 116, 553, 1980.

121. **Ishiguro, N. and Sato, G.,** The distribution of plasmids determining citrate-utilization in citrate-positive variants of *Escherichia coli* from humans, domestic animals, feral birds and environments, *J. Hyg.,* 83, 331, 1979.

122. **Reynolds, C. H. and Silver, S.,** Citrate utilization by *Escherichia coli:* plasmid- and chromosome-encoded systems, *J. Bacteriol.,* 156, 1019, 1983.

123. **Ishiguro, N., Sato, G., Sasakawa, C., Danbara, H., and Yoshikawa, M.,** Identification of citrate utilization transposon Tn3411 from a naturally-occurring citrate utilization plasmid, *J. Bacteriol.,* 149, 961, 1982.

124. **Shinagawa, M., Makino, S., Hirato, T., Ishiguro, N., and Sato, G.,** Comparison of DNA sequences required for the function of citrate utilization among different citrate utilization plasmids, *J. Bacteriol.,* 151, 1046, 1982.

125. **Summers, A. O., Jacoby, G. A., Swartz, M. N., McHugh, G., and Sutton, H.,** Metal cation and oxyanion resistances in plasmids of Gram negative bacteria, in *Microbiology — 1978,* Schlessinger, D., Eds., American Society of Microbiology, Washington, D.C., 1978, 128.

126. **Schottel, J. L.,** The mercuric and organomercurial detoxifying enzymes from a plasmid-bearing strain of *Escherichia coli, J. Biol. Chem.,* 253, 4341, 1978.

127. **Tezuka, T. and Tonomura, K.,** Purification and properties of a second enzyme catalysing the splitting of carbon mercury linkages from mercury-resistant *Pseudomonas* K-62, *J. Bacteriol.,* 135, 138, 1978.

128. **Schottel, J., Mandal, A., Clark, D., Silver, S., and Hedges, R. W.,** Volatilization of mercury and organomercurials determined by inducible R-factor systems in enteric bacteria, *Nature (London),* 251, 335, 1974.

129. **Weiss, A. A., Murphy, S. D., and Silver, S.,** Mercury and organomercurial resistances determined by plasmids in *Staphylococcus aureus, J. Bacteriol.,* 182, 197, 1977.

130. **Weiss, A. A., Schottel, J. L., Clark, D. L., Beller, R. G., and Silver, S.,** Mercury and organomercurial resistance with enteric, staphylococcal and pseudomonad plasmids, in *Microbiology — 1978,* Schlessinger, D., Ed., American Society of Microbiology, Washington, D.C., 1978, 121.

131. **Summers, A. O., Weiss, R. B., and Jacoby, G. A.,** Transposition of mercury resistance from a transferable R plasmid of *Escherichia coli, Plasmid,* 3, 35, 1980.

132. **Summers, A. O. and Jacoby, G. A.,** Plasmid determined resistance to tellurium compounds, *J. Bacteriol.,* 129, 276, 1977.

133. **Silver, S., Budd, K., Leahy, K. M., Shaw, W. V., Hammond, D., Novick, R. P., Willsky, G. R., Malamy, M. H., and Rosenberg, H.,** Inducible plasmid-determined resistance to arsenate, arsenite and antimony (III) in *Escherichia coli* and *Staphylococcus aureus, J. Bacteriol.,* 146, 983, 1981.

134. **Silver, S. and Keach, D.,** Energy-dependent arsenate efflux: the mechanism of plasmid-mediated resistance, *Proc. Natl. Acad. Sci. U.S.A.,* 79, 6114, 1982.

135. **Mobley, H. L. T. and Rosen, B. P.,** Energetics of plasmid-mediated arsenate resistance, in *Escherichia coli, Proc. Natl. Acad. Sci. U.S.A.,* 79, 6119, 1982.

136. **Rodriguez-Lemoine, V.,** A simple method for the detection of conjugative plasmids of the incompatibility group H2, *Microb. Lett.,* 21, 35, 1982.

137. **Shaw, W. V.,** Chloramphenicol acetyltransferase: enzymology and molecular biology, *CRC Crit. Rev. Biochem.,* 14, 1, 1983.

138. **Foster, T. J.,** Plasmid-determined resistance to antimicrobial drugs and toxic metal ions in bacteria, *Microbiol. Rev.,* 47, 361, 1983.

139. **Smith, A. L. and Burns, J. L.,** Resistance to chloramphenicol and fusidic acid, in *Antimicrobial Drug Resistance,* Bryan, L. E., Ed., Academic Press, New York, 1984, 293.

140. **Fitton, J. E. Packman, L. C., Harford, S., Zaidenzaig, Y., and Shaw, W. V.,** Plasmids and the evolution of chloramphenicol resistance, in *Microbiology — 1978,* Schlessinger, D., Ed., American Society of Microbiology, Washington, D.C., 1978, 249.

141. **Foster, T. J. and Shaw, W. V.,** Chloramphenicol acetyltransferase specified by fi⁻ R-factors, *Antimicrob. Agents Chemother.,* 3, 99, 1973.

142. **Gaffney, D. F., Foster, T. J., and Shaw, W. V.,** Chloramphenicol acetyltransferases determined by R plasmids from Gram-negative bacteria, *J. Gen. Microbiol.,* 109, 351, 1978.

143. **Maher, D. and Colleran, E.,** The chloramphenicol resistance determinant of HI2 plasmids, *Biotech. Lett.,* (In press).

144. **Volker, T. A., Iida, S., and Bickle, T. A.,** A single gene coding for resistance to both fusidic acid and chloramphenicol, *J. Mol. Biol.,* 154, 417, 1982.

145. **Iida, S.,** On the origin of the chloramphenicol resistance transposon Tn9, *J. Gen. Microbiol.,* 129, 1217, 1983.

146. **Terakado, N., Sekizaki, T., Hashimoto, K., Yamagata, S., and Yamamoto, T.,** Chloramphenicol transposons found in *Salmonella naestved* and *Escherichia coli* of domestic animal origin, *Antimicrob. Agents Chemother.,* 20, 382, 1981.

147. **Alfaro, G.,** R plasmids from *S. typhi* and *S. typhimurium* strains isolated in Mexico City hospitals, in *Molecular Biology, Pathogenicity and Ecology of Bacterial Plasmids,* Levy, S. B., Clowes, R. C., and Koenig, E. L., Eds., Plenum Press, New York, 1981, 575.

148. **Gangarosa, E. J., Bennett, J. V., Wyatt, C., Pierce, P. E., Olarte, J., Hernandes, P. M., Vazquez, V., and Bessudo, D.,** An epidemic-associated episome?, *J. Infect. Dis.,* 126, 215, 1972.

149. **Butler, T., Shuster, C. W., and Franco, A.,** Effect of a *Salmonella* group H1 R factor on virulence and response of infections to antimicrobial therapy, *Antimicrob. Agents Chemother.,* 15, 478, 1979.

150. **Datta, N., Richards, H., and Datta, C.,** *Salmonella typhi in vivo* acquires resistance to both chloramphenicol and co-trimoxazole, *Lancet,* i, 1181, 1981.

151. **Timoney, J. F. and Linton, A. H.,** Experimental ecological studies on H2 plasmids in the intestine and faeces of the calf, *J. Appl. Bact.,* 52, 417, 1982.

152. **Linton, A. H.,** R-plasmids in microorganisms found on food, in *Resistance and Pathogenic Plasmids,* Proc. of CEC Seminar, Brussels, 15—16 October, 1981, Pohl, P. and Leunen, J., Eds., National Institute for Veterinary Research, Brussels, 75, 1981.

153. **Cooke, E. M., Shooter, R. A., Kumar, P. J., Rousseau, S. A., and Foulkes, A. L.,** Hospital food as a possible source of *Escherichia coli* in patients, *Lancet,* i, 436, 1970.

154. **Linton, A. H., Howe, K., Hartley, C. L., Clements, H. M., and Richmond, M. H.,** Antibiotic resistance among *Escherichia coli* O-serotypes from the gut and carcases of commercially slaughtered broiler chickens: a potential public health hazard, *J. Appl. Bact.,* 42, 363, 1977.

155. **Cooke, E. M., Hettiaratchy, I. G. T., and Buck, A. C.,** Fate of ingested *Escherichia coli* in normal persons, *J. Med. Microbiol.,* 5, 361, 1972.

156. **Linton, A. H., Howe, K., Bennett, P. M., Richards, M. H., and Whiteside, E. J.,** The colonization of the human gut by antibiotic resistant *Escherichia coli* from chickens, *J. Appl. Bact.,* 43, 465, 1977.

157. **Hartley, C. L., Clements, H. M., and Linton, K. B.,** *Escherichia coli* in the faecal flora of man, *J. Appl. Bact.,* 43, 261, 1977.

158. **Broda, P.,** *Plasmids,* W. H. Freeman, Oxford, 1979, chap. 6.

159. **Burman, L.,** Expression of R-plasmid functions during anaerobic growth of an *Escherichia coli* K-12 host, *J. Bacteriol.,* 131, 69, 1977.

160. **Singleton, P. and Anson, A. E.,** Effect of pH on conjugal transfer at low temperatures, *Appl. Environ. Microbiol.,* 46, 291, 1983.

161. **Anderson, J. D., Ingram, L. C., Richmond, M. H., and Wiedemann, B.,** Studies on the nature of plasmids arising from conjugation in the human gastro-intestinal tract, *J. Med. Microbiol.*, 6, 475, 1973.
162. **Petrocheilou, V., Grinsted, J., and Richmond, M. H.,** R-plasmid transfer *in vivo* in the absence of antibiotic selection pressure, *Antimicrob. Agents Chemother.*, 10, 753, 1976.
163. **Smith, M. G.,** *In vivo* transfer of an R-factor within the lower gastrointestinal tract of sheep, *J. Hyg.*, 79, 259, 1977.
164. **Jones, R. T. and Curtiss, R., III,** Genetic exchange between *Escherichia coli* strains in the mouse intestine, *J. Bacteriol.*, 103, 71, 1970.
165. **Lafont, J.-P., Bree, A., and Plat, M.,** Bacterial conjugation in the digestive tracts of gnotoxenic chickens, *Appl. Environ. Microbiol.*, 47, 639, 1984.
166. **Lafont, J.-P., Guillot, J. F., Chaslus-Dancla, E., Dho, M., and Eucher-Lahon, M.,** Antibiotic-resistant bacteria in animal wastes: a human health hazard, *Bull. Inst. Pasteur (Paris)*, 79, 213, 1981.
167. **Fontaine, T. D., III and Hoadley, A. W.,** Transferable drug resistance associated with coliforms isolated from hospital and domestic sewage, *Health Lab. Sci.*, 13, 238, 1976.
168. **Bell, R. B.,** Antibiotic resistance patterns of faecal coliforms isolated from domestic sewage before and after treatment in an aerobic lagoon, *Can. J. Microbiol.*, 24, 886, 1978.
169. **Grabow, W. O. K., Von Zyl, M., and Prozesky, O. W.,** Behaviour in conventional sewage purification processes of coliform bacteria with transferable and non-transferable drug resistance, *Water Res.*, 10, 717, 1976.
170. **Mach, P. A. and Grimes, D. J.,** R-plasmid transfer in a wastewater treatment plant, *Appl. Environ. Microbiol.*, 44, 1395, 1982.
171. **Grabow, W. O. K., Prozesky, O. W., and Burger, J. S.,** Behaviour in a river and dam of coliform bacteria with transferable or non-transferable drug resistance, *Water Res.*, 9, 777, 1975.
172. **Petrocheilou, V., Richmond, M. H., and Bennett, P. M.,** Spread of a single plasmid clone to an untreated individual from a person receiving prolonged tetracycline therapy, *Antimicrob. Agents Chemother.*, 12, 219, 1977.
173. **Taylor, D. E.,** personal communication.
174. **Maher, D. and Colleran, E.,** unpublished.
175. **Bradley, D. E.,** The unique conjugation system of Inc HI3 plasmid MIP233, *Plasmid*, 16, 63, 1986.

Chapter 2

APPLICATION OF BIOTECHNOLOGY TO DAIRY INDUSTRY WASTE

Vidar F. Larsen and Ian S. Maddox

TABLE OF CONTENTS

I. INTRODUCTION

Wastes from dairy factories can be divided into two categories: liquid by-products and wash waters. Included in the former is whey, the fluid obtained by separating the coagulum from milk, cream, or skimmed milk. Although the wash waters from the plant can be of considerable volume, the whey contains by far the largest fraction of fermentable carbohydrate. It is therefore this fraction of the dairy industry wastes which can be used most successfully, and it is this fraction which also causes considerable environmental problems today.

Whey is produced in the manufacture of cheese, rennet, and acid casein, and each process gives rise to a characteristic whey. The annual production of whey is relatively large, with the five largest dairy producers disposing of or processing nearly 50,000 tonnes, the majority of which is sweet whey (Table 1). Sweet whey, pH > 5.6, is derived from the manufacture of cheese or rennet casein, and it is produced on average at a rate of 7.6 kg/kg of cheese. Acidic whey, pH < 5.1, is obtained from processes where acid is either added to the process or produced in the process to facilitate coagulation. Typical processes are lactic or sulfuric casein, and some 25 kg of whey is produced per kilogram of casein.[1]

The volume and strength of the waste from dairy factories is considerable, and unless these wastes can be further utilized, they must be treated prior to disposal. Land disposal of such wastewaters is a recognized means of pollution control which also can increase soil fertility and augment rainfall. In New Zealand, approximately one half of the dairy factories use this method of waste disposal.[2] Despite the use of such systems for decades, problems may occur in their operation. These problems are mainly due to poor initial design, bad construction, and/or incorrect operation. One of the more common mistakes in the operation is to overload the land, giving rise to water saturation or burning of the soil by the strong waste. Aerobic waste treatment methods are also used successfully for the treatment of such waste. There is a considerable volume of literature on this subject, and it is not the intention of the authors to review this here.

Considerable effort has been made over the past few years to recover protein from whey. Ultrafiltration plants are being installed in the U.S., Europe, and New Zealand for the

Table 1
ESTIMATED PRODUCTION OF WHEY IN 1977[1]

Country	Sweet whey		Acid whey		Total	
	Liquid	**Solids**	**Liquid**	**Solids**	**Liquid**	**Solids**
U.S.	13,720	892	1,920	125	15,640	1,017
Canada	1,210	79	158	10	1,368	89
EEC	22,800	nq	nq	nq	~25,000	1,490
Australia	788	51	420	27	1,208	79
New Zealand	785	51	1,282	53	2,067	131

Note: Data for EEC are for 1977. Units in thousands of tonnes; nq = not quoted.

Table 2
TYPICAL COMPOSITION OF WHEYS
PRODUCED IN NEW ZEALAND[4]

	Cheese whey(g/kg)	Casein wheys (g/kg)	
		Lactic	**Sulfuric**
Total solids	67	64	59
Lactose	50	44	47
Protein	5.7	5.7	5.3
Nonprotein nitrogen	0.5	0.5	0.3
Fat	0.3	0.2	0.4
Ash	5.3	5.8	6.7

production of whey protein concentrates.[3] The resultant recent interest in the application of biotechnological processes has also been applied to whey permeate.

II. WHEY COMPOSITION

The composition of whey varies according to the process from which it derives. It usually contains half the initial total solids (TS) content of the incoming milk, and averages about 6% of the TS. The major constituents of the whey solids are lactose and protein (Table 2), which should make whey an excellent fermentation medium. The main problem is the sugar concentration of whey. Although this is of a sufficiently high level for yeast production, it is well below that considered optimum for fermentations like ethanol and citric acid.

If whey permeate is to be used, the lactose composition is further reduced, and the protein concentration is virtually zero (Table 3). Although the availability of some minerals is high, supplementation of some other minerals may be necessary. It can be seen from Table 4 that the supply of cobalt, iron, magnesium, and phosphorous is very low, as is the supply of manganese when high growth rates and cell concentrations are required.

III. LACTOSE MANUFACTURE

The technology for the manufacture of crystalline lactose monohydrate from whey has been well described,[8,9] the process involving concentration by evaporation, crystallization, separation, refining, drying, and milling. In some respects, protein-free whey is a better starting material for lactose manufacture since the presence of whey proteins limits the degree

Table 3
COMPOSITION OF DEPROTEINATED WHEYS

	Cheddar[5] cheese	Cottage[5] cheese	Lactic[3] casein	Sulfuric[6] casein
Total solids	57.0	58.0	56.8	56.4
Lactose	49.0	43.0	44.8	46.0
Total nitrogen	0.26	0.33	0.64	0.37
Nonprotein nitrogen	0.24	0.30	0.46	0.32
Ash	5.0	5.6	5.7	7.9
pH	6.1	4.7	nq	5.0

Note: Units are in grams per kilogram; nq = not quoted.

Table 4
MINERAL CONTENT OF LACTIC
WHEY PERMEATE[7]

Metal	Concentration (mg/ℓ)
Aluminium	5.1
Calcium	835.1
Cobolt	—
Copper	0.3
Iron	1.2
Potassium	881.6
Magnesium	62.6
Manganese	0.04
Sodium	290.2
Nickel	—
Phosphorous	382.6
Sulfur	43.1
Strontium	0.45
Zinc	43.5

Note: The dash indicates that the content was below the detection limit.

of evaporation and impedes crystallization and separation.[6] Conversely, deproteinated whey is often virtually saturated with calcium ions, so that precipitation of calcium salts occurs during evaporation resulting in fouling of heat exchange surfaces and contamination of lactose crystals.[9] Therefore, it is generally accepted that deproteinated whey should be pretreated prior to evaporation to reduce its ionic content. Suitable processes include ion exchange and electrodialysis, both of which have been well reviewed by Short.[6] The manufacture of crystalline lactose from lactic acid casein whey (permeate) is further complicated by the presence of relatively high concentrations of lactate. An interesting approach to the solution of this problem has been described by Ruiz et al.,[10] who treated the whey with the yeast *Candida ingens* to remove the lactate. The process removed 98% of the lactic acid and 40% of the nonprotein nitrogen from the whey, while having no effect on the lactose. During the fermentation process the pH rose from 4.4 to 8.0, causing precipitation of calcium salts which could be removed, together with the yeast, by filtration or centrifugation. The resulting whey could be further processed without complications. Another approach to recover lactose from lactic casein whey permeate has been described by Hobman.[9] The process involves partial removal of calcium phosphate complexes prior to evaporation using an alkali and heat treatment precipitation, followed by centrifugal clarification. A variety of alkalies were investigated, including sodium hydroxide, calcium hydroxide, and sodium carbonate, either alone or in combination. Subsequent pilot-scale trials revealed that removal of approximately 50% of the calcium was sufficient to avoid difficulties during evaporation.

At present, the market for lactose appears to be limited,[6,9] since its use in the food industry suffers from its properties of low solubility and lack of sweetening power. However, it does have the advantage of conferring flavor enhancement and protein stability. Nevertheless, Hobman has reported that since the capital and/or operating costs of many of the pretreatment processes essential for the manufacture of lactose from protein-free whey may be prohibitively large, the profitability of the process is very dependent on economies of scale and use of an energy- and yield-efficient process.[9] On this basis, then, development of alternative processes for lactose recovery may be warranted. One such process is the Steffen process, which makes use of the ability of carbohydrates to complex with alkaline-earth metals, and forms the basis of sucrose recovery from sugar beet molasses. Its application to lactose solutions has been described by several groups,[11-13] and lactose recoveries of up to 97% have been reported from 5% (w/v) aqueous solutions.[13]

IV. LACTOSE HYDROLYSIS

The hydrolysis of lactose using β-galactosidase (lactase) gives rise to equimolar amounts of glucose and galactose. Such a process markedly improves the two properties of greatest commercial importance in sugars, i.e., sweetness and solubility, and thus greatly increases the range of potential applications of the product in the food industry.[14-16] The uses envisaged for such products are as substitutes for corn syrup or, in some cases, sucrose.[15] However, for many of these applications it is clear that demineralized products must be used. Ennis has summarized the various types of products and their applications:[16]

1. Undemineralized hydrolyzed permeate syrups; use as fermentation substrates and in animal feeds.
2. Demineralized (50, 70, 90%) hydrolyzed permeate syrups; use in a range of foods where sucrose and/or glucose syrups are traditionally used, e.g., ice cream, confectionery, beer brewing, and soft drinks.
3. Undemineralized hydrolyzed whey syrups; use in food products where salty taste is acceptable, e.g., chocolate drinks.
4. Partially or totally demineralized hydrolyzed whey syrups; use in bakery products and dairy desserts.

β-Galactosidases are widely distributed in nature, but for dairy industry application, preparations from yeasts *(Kluyveromyces lactis, K. fragilis)* and fungi *(Aspergillus niger, A. oryzae)* are the ones of major interest,[17-19] and are available commercially. The enzyme from *Escherichia coli* is also available commercially, but finds little application in the treatment of whey. The fungal enzymes generally have acidic pH optima in the range 2.5 to 5.0, whereas yeast lactases are more active in the range 6.0 to 7.0. Thus, the fungal enzymes are used mainly for the treatment of acid wheys, and the yeast enzymes for sweet wheys and milk. With regard to their temperature optima, fungal enzymes are more heat stable and can be used at 55°C, whereas the optimum for the yeast enzymes is about 35°C. It would be advantageous, therefore, if enzymes could be found which operate at netural pH and have a higher heat stability than those obtained from yeast. Ramana Rao and Dutta have described a β-galactosidase from *Streptococcus thermophilus* which can be used at 45°C and pH 7.0, and is more stable on storage than the yeast enzyme.[20] Also, a thermostable enzyme from a thermophilic *Bacillus,* showing optima at 60°C and pH 6.0, has been reported by Griffiths and Muir,[21] and a preparation from the yeast *C. pseudotropicalis* has been shown to have optima at 47°C and pH 6.2.[22] These enzymes may have potential in the commercial hydrolysis of lactose, as might preparations from various species of *Lactobacillus,* although they have been less well characterized.[23]

The simplest way to achieve lactose hydrolysis is to directly add the β-galactosidase enzyme to the whey or whey permeate, and to hold at the required temperature and pH until the desired extent of hydrolysis is attained. This, however, is usually uneconomic since a large amount of enzyme is required. To overcome this, recovery and reuse of the enzyme using, e.g., ultrafiltration, can be practiced.[9,19] A more satisfactory process, however, is to immobilize the enzyme to a solid support so that the lactose solution is brought into intimate contact with the enzyme which can subsequently be easily separated after hydrolysis is complete. A whole variety of immobilization methods can be employed, and these have been the subject of recent reviews.[15,18,23,24] Although enzyme immobilization may reduce the high operating costs which are associated with ''free'' enzyme processes, the large capital investment required generally makes immobilized enzyme processes dependent on economies of scale.[9] Aspects of the design and operation of various immobilized enzyme reactor types have been reviewed by Pitcher,[25] while some pilot-plant processes for the hydrolysis of whey (permeate) have been described by Coton,[15] Ennis,[16] Prenosil et al.,[17] and Dicker.[26] Enzyme processes, however, do suffer from some disadvantages, including loss of enzyme activity over time, practical limits of about 80% hydrolysis due to product inhibition, the production of a small quantity of indigestible oligosaccharides due to transferase activity of the enzyme, and the problem of maintaining sterility in the reactor.[6] The yeast enzymes also have a requirement for some metal ions to maintain their activity, and this is a distinct disadvantage in the food industry.[23]

An alternative process is acid hydrolysis. This method does not suffer from product inhibition or microbial contamination but can generally only be used with deproteinated whey since the presence of proteins causes unacceptably high levels of off-flavor and color.[15,19] However, decolorization and flavor removal may be performed using activated carbon to produce a water-white syrup.[6] The formation of oligosaccharides appears to be less during acid hydrolysis than enzymatic hydrolysis.[19] ''Homogeneous'' or ''single phase'' acid hydrolysis is performed using hydrogen ions in solution, e.g., pH 1.0 to 1.5, during a defined heat treatment, e.g., 60°C for 24 hr or 140°C for 11 min.[9] The hydrogen ions may be provided by either direct acidification with mineral acids or exchange of hydrogen ions for cations in solution using ion-exchange resins. ''Heterogeneous'' or ''two-phase'' processes employ ''insoluble'' hydrogen ions, bound to a cation-exchange resin, to catalyze the hydrolysis.[9,27] Here, the whey permeate is completely decationized, heated to 90 to 98°C, and then passed through a cation-exchange resin (in the H^+ form) at a flow rate sufficient to provide the residence time (80 min) required for hydrolysis. The product is finally passed through an ion-exchange resin to remove anions, color, and off-flavors. The ''heterogeneous'' system is claimed to be more cost-effective than the homogeneous system.[27]

The profitability of manufacturing hydrolyzed lactose syrups for use in the food industry is largely dependent on the price of competitive products such as sucrose and corn syrups. Hobman has commented that since there has been no real price increase for sugar over the last decade, and because of the increasing capital investment required for a new plant, the production of hydrolyzed lactose syrups has become uneconomic.[9] However, further treatment of the syrup with the enzyme glucose isomerase, to increase the sweetness, could improve price competitiveness.[28]

V. ANAEROBIC DIGESTION

The aerobic waste treatment methods briefly mentioned above are capable of giving a high degree of waste stabilization when applied to dairy wastes. They are, however, perhaps with the exception of spray irrigation, known to be costly when treating large quantities of wastewater. Furthermore, they produce large quantities of biological solids which must be further processed. The cost of handling these solids contributes considerably to the overall waste treatment cost.

Anaerobic waste treatment offers an alternative and beneficial method of handling dairy wastes. Due to the much lower production of ATP under anaerobic metabolism compared with aerobic metabolism, much lower energy is available to the organisms for cell synthesis. The yield of biological solids is therefore low, as is the need for further solids handling. The process has further advantages over an aerobic one, in that:

1. It has no requirement for oxygen.
2. It produces a useful end product in the form of methane.
3. It has a lower requirement for additional nutrients.

During the digestion process the organic material is converted to methane and carbon dioxide. For well-designed digester systems, in excess of 90% of the influent carbon can end up in the gas phase.

The main disadvantage with anaerobic processes has been the long hydraulic residence time required in order to obtain a high degree of stabilization. This meant that large digester volumes were required, and led to large capital investments. With the onset of the energy crisis of the 1970s, the interest in the process took a new turn, and the advances in the understanding of the basic principles and in the engineering design have been great. This had led to improved process stability and economics.

Anaerobic digestion of whey was tried as early as the 1930s when Buswell and co-workers[29] showed that a vast variety of dairy industry wastes could be treated by anaerobic digestion.

A. Anaerobic Lagoon

The simplest design of anaerobic digestion systems is the lagoon. The operation of such a system is usually at ambient temperature, and therefore it operates under reduced metabolic activity, particularly during the winter months. Investigating the reduction of BOD_5 in a 30-ft^2, 5-ft deep lagoon, Parker and Skerry found that 99% of the BOD_5 could be removed at organic loading rates of 445 kg of BOD per hectare per day.[30] Although such operations are feasible, it can only be used under conditions where there are large areas of land available at low cost, as the treatment of 1000 m^3 whey per day with a BOD_5 of 36 kd/m^3 would require in excess of 80 ha.

The treatment of dairy industry waste, in combination with other food industry wastes and domestic waste, has also been shown to be successful in lagoons.[31] Combining such wastes in full-scale lagoons, it was found that during the spring period, when the dairy season was at a peak, BOD_5 reductions of 85% could be achieved. During this period dairy waste accounted for 36% of the BOD_5 load and the loading rate varied between 124 and 246 kg of BOD per hectare per day. Improved performance could be achieved when additions of nitrogen were made to the waste to keep the carbon to nitrogen ratio below 40. Under such conditions a loading rate of 854 kg/ha/day could be achieved with a BOD_5 reduction of 88 to 89%.

B. Stirred Tanks

The use of lagoons as anaerobic digesters often makes it impossible to take advantage of the main benefit of the process, namely harvesting of the methane. For this reason, tank digesters are most frequently used, and the conventional systems usually consist of circular concrete tanks with fixed or floating covers. Feed was on a continuous or intermittent basis, and hydraulic retention times of 30 to 60 days were not uncommon. Improved digestion rates were obtained by the introduction of mixing, either by mechanical means, or by gas recirculation. Such mixing, together with the introduction of temperature control, saw the retention times reduce to between 10 and 30 days.

Using a laboratory-scale digester, Follmann and Markl showed that short retention times could be achieved with whey digesters.[32] Using pH control to maintain the pH between 6.95 and 7.0, they showed that 98% COD reduction was possible with a retention time of 12.5 days. This gave an organic loading rate of 4 kg of COD per cubic meter per day. The digester was operated in the thermophilic range with a temperature of 55°C. The gas production from the digester was 3 m^3/m^3/day with a methane content of 50%.

C. Contact Digester

When dealing with dilute wastes there is always a problem with tank digesters in that the biomass occurring within them is too small to give fast digestion times and fast response to process upsets. The anaerobic contact process was developed in response to this need to maintain high concentrations of active biomass in the digester. This configuration is analogous to the activated sludge process, and was first applied to the treatment on meat packing wastes. In such a system the biomass in the digester may reach 5 to 10 g volatile suspended solids (VSS) per liter of digester volume, and typical loading rates are 5 to 10 kg of COD per cubic meter per day.

The main problem with such a digester system is associated with the solids separation step. Degasification of the sludge or the use of flocculants often is required when dealing with soluble waste. There is a further requirement for external heating when dealing with dilute waste as such waste generates insufficient methane to maintain a 35°C operating temperature. The contact process has been evaluated on dilute dairy waste.[34] Using a 80-m^3 digester with a 19.5-m^3 settling tank, waste containing between 4.2 and 5.0 kg of COD per cubic meter was treated with 92 to 96% BOD_5 removal. The digester hydraulic retention time was 1.89 days, but if the settling tank and associated pipework was taken into account, it was in excess of 2.4 days. Gas production was between 1.06 and 1.20 m^3/m^3/day, when basing it on the digester volume alone with a methane content of approximately 75%. This system used the so-called thermal shock process to improve the settlement of the sludge. Sludge from the digester was passed through an external heat exchanger to reduce the temperature from 35 to 25°C before being passed to the settling tank.

Dilute dairy waste has also been treated in a combination of the anaerobic contact process and activated sludge plant.[35] Using a 15- to 25-m^3 anaerobic stirred tank, and a 2.5- to 4.0-m^3 aerobic tank, BOD_5 removals of 98% were achieved. The organic loading rate in the digester was 2.0 kg COD per cubic meter per day, and gas production was 0.45 m^3/m^3/day with a methane content of 85%. No details were given for the sludge separation system.

A modification of the contact process was proposed by Callander and co-workers.[36] Using an intermittently stirred tank digester and addition of flocculant to the digester itself, they obtained high biomass retention in the digester. The biomass concentration obtained varied from 35 to 56 kg/m^3 of VSS, which compared with 19.1 kg/m^3 obtained in a digester system without flocculant. This high biomass level ensured that 99% of the soluble COD was removed at loading rates of 12 to 16 kg COD per cubic meter per day. Treating full-strength whey with a COD of 69.8 kg/m^3 gave a hydraulic retention time of between 4.4 and 5.8 days. This compares favorably with the minimum of 7.7 days obtained for the digester without flocculant. During operation the pH of the feed was adjusted to 6.5 to maintain a digester pH of between 6.8 and 7.0.

The addition of flocculant was intermittent, and the average concentration was 1.6 m^3 of 5% flocculant per 1000 m^3 of digester volume. The gas production was between 5 and 8 m^3/m^3/day, with a methane content of 50 to 55%.

D. The Upflow Anaerobic Sludge Blanket (UASB)

This type of digester system can be said to be a natural extension of this contact process. It is based on the existence of naturally occurring flocculant microorganisms. By a careful and controlled start-up procedure, flocs of up to 5 mm in size may be obtained. These flocs are of sufficient size and density to settle against the upflow of liquid. Thus a bed of microorganisms are obtained in the bottom of the digester, with a concentration reaching 60 kg/m^3 of VSS.[33]

The application of the UASB to the treatment of whey has been studied by Archer et al.[7,37] Initial operation of the UASB digester was limited to diluted lactic casein whey permeate at a strength of 60% of normal permeate. At this strength BOD$_5$ and COD removal efficiencies in excess of 99% were obtained at a hydraulic retention time of 12.4 days. When the initial problems with retention of the biomass in the digester had been overcome, full-strength whey permeate, pH 4.7, was treated successfully in the digester. At a retention time of 11 days (organic loading rate or 4.3 kg/m^3/day), COD reduction was 98.2% and BOD$_5$ reudction was 97.5%.[37] Gas production was 2.6 m^3/m^3/day with a methane content of 54%. The digester was operated with a recycle to provide natural buffering capacity to the system, as no pH adjustment was made to the feed. With a recycle rate of approximately one digester volume per 40 hr, the pH in the digester was maintained at 7.0 or above. The flow regime of the digester approached that of a stirred tank, and only small COD and biomass profiles were observed over the length of the digester. When treating diluted whey, the soluble COD decreased to less than 0.5 kg/m^3, with an effluent COD of 0.43 kg/m^3. The VSS was nearly constant at all sampling points at approximately 20 to 25 kg/m^3 (Figure 1). Increased organic loading appears to have only a small effect on the COD reduction efficiency. Operating the UASB at a loading rate of 20.9 kg COD per cubic meter per day, the COD reduction decreased only to 97.9%.[38] At this retention time, 2.4 days, gas production was 14.3 m^3/m^3/day with a methane content of 51%.

Using the UASB on diluted dairy waste, loading rates of up to 15 kg/m^3/day of COD could be achieved.[39] The total COD reduction was, however, found to be strongly dependent on the loading rate, with 87 and 73% reduction occurring at 5 and 15 kg/m^3/day, respectively. The suspended biomass was found to account for approximatley 35% of the total effluent COD. The methane production rate was between 1 and 3 m^3/m^3/day.

E. Anaerobic Filter

In an effort to retain the microorganisms in the digester, several investigators have used inert material in the tank. This material provides a surface to which the organisms can attach and thus not be washed out with the effluent liquid. The digester therefore closely resembles the aerobic trickling filter. A variety of inert materials have been used, ranging from rock fillings with low voidage to plastic media of high voidage. The filter has also been tested for a range of wastes, ranging from synthetic media based on glucose or acetic acid to pharmaceutical and petrochemical wastes. Direct comparison between published filter data is difficult as it has been shown that the particular filter design used greatly influences filter performance.[40]

Although the ground-breaking work on the filter was performed by Young and McCarty,[41] the method was used as early as the 1930s. Buswell et al.[29] used laboratory digesters partly filled with asbestos fibers to successfully treat a variety of dairy wastes at temperatures between 27 and 29°C. Using whey with a BOD$_5$ of 35 kg/m^3 and a pH of 4.9, a BOD$_5$ reduction of 93% was obtained at a hydraulic retention time of 29 days. Gas production was 1.6 m^3/m^3/day, with a methane content of 49.8%. The investigators found little difference when diluting the whey with an equal part of water, or when operating the digester in the thermophilic range (53 to 58°C). Other wastes were successfully treated with skimmed milk and buttermilk/whey mixtures, either undiluted or diluted.

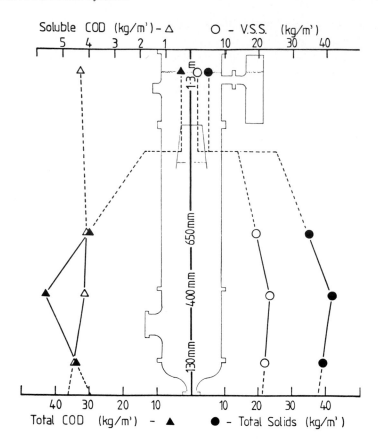

FIGURE 1.   Profile of the sludge blanket.[37]

An attempt to treat diluted cheese whey (COD 1.8 to 8.1 kg/m³) and dilute whey permeate (COD 4.1 to 5.0 kg/m³), was made using rock-filled anaerobic filters.[42] Although good treatment efficiencies were obtained for short periods with the filter-treating whey, it was found necessary to supplement both digesters with 1.5 g/kg nutrient broth in order to maintain stable operation. The performance of the whey filter was superior to that of the whey permeate filter. The former was loaded at the maximum at a rate of 1.8 to 1.9 kg COD per cubic meter per day a hydraulic retention time of 4.2 days. At this rate, COD reduction was 97%. The maximum loading rate obtained with whey permeate was 0.5 kg/m³/day. The maximum loading rate obtained with whey permeate was 0.5 kg/m³/day. The reason for this apparent failure to treat whey on its own was stated as a lack of essential micronutrient(s) in the whey, although the filter design used did not employ a effluent recycle or pH control.

In an attempt to overcome the instability observed with a one-pass filter when treating acidic waste, Maranan investigated the treatment of lactic casein whey in a filter with effluent recycle.[43] Employing a feed to recycle flow ratio of 1:20, stable operation was obtained at an organic loading rate of 3.6 kg COD per cubic meter per day and a hydraulic retention time of 20 days. At this retention time the COD and the BOD_5 reductions were 88 and 91%, respectively, and gas production was 1.1 m³/m³ with a methane content of 60%. It was, however, necessary to add alkalinity to the feed at intervals to maintain stable operation. At approximately 5- to 7-day intervals, 40 to 50 mℓ of 5 *N* NaOH were pumped into the 7-ℓ filter. These additions ensured that the filter pH was maintained above 7.1. An increase in loading rate to 4.8 kg/m³/day saw a decrease in the treatment efficiencies. The COD reduction was reduced to 82% and the BOD_5 to 80%. No change was observed in the gas production or in the methane content.

In an attempt to investigate the effect of process failure in the form of recycle blockage, the recycle was removed abruptly when operating at 20 days retention time. Within 5 days the COD reduction had decreased to 55%, and the gas production and methane concentration had reduced to half their previous values. During this period the pH of the feed was maintained at 7.0. A small recovery was noted over a 10-day period, and a "steady-state" condition was observed over the following 26 days. The COD and BOD_5 reductions during this period were 57 and 63%, respectively. The reintroduction of the recycle did not improve the treatment efficiencies.

A variation on the traditional filter design was developed at the National Research Council of Canada. In this design the biomass is attached to fixed surfaces in the digester. The support material forms channels of relatively small cross-sectional areas through which the waste flows downwards. As all the biomass is attached to the surface, this digester design is not suceptible to blocking and channeling as the traditional filter is, and it has been found that it is able to withstand severe hydraulic shock loads.[39] Using this design to treat full-strength whey (COD 66 kg/m^3), total COD reductions of 94% were obtained at an organic loading rate of 22 $kd/m^3/day$. Increased removal efficiencies were found as the loading rate decreased. The methane production rate was between 3 and 6 $m^3/m^3/day$. The investigator found no benefit in using effluent recycle or adjusting the pH prior to feeding the digester.

When treating dilute dairy industry waste (COD 4.0 kg/m^3), it was found, however, that the loading rate had to be limited to 15 $kg/m^3/day$, and that the treatment efficiency was less than that for full-strength whey. The total COD reduction at 15 $kg/m^3/day$ was 73% with a suspended COD of approximately 35% of the effluent total COD. Using the same digester design on diluted cheese whey, de Haast et al.[44] found that COD reductions of between 85 and 87% were possible at organic loading rates of 3.1 to 3.8 kg COD per cubic meter per day. This gave hydraulic retention times of between 4.2 and 3.5 days.

F. Expanded and Fluidized Beds

The main operational problems experienced with the traditional anaerobic filter (packed bed reactor) is its tendency to block suspended material, if any is present, and to channel the liquid. If the packing material is descrete particles, the voidage between them can be increased by increasing the liquid flow rate up the digester. This will overcome these problem tendencies even if small suspended particles are present. The increased flow rate is usually obtained by the introduction of effluent recycle.

When increasing the flow rate above that of the packed bed, a rate will be reached when the bed begins to expand; thus the voidage is increased. A limiting flow rate will be reached when the solids particles are no longer supported physically by the digester; the bed is fluidized. At this point the hindered settling velocity of the particles matches that of the liquid velocity in the digester. Careful consideration must be given to match these velocities to obtain fluidization at the bottom of the digester, where the voidage will be low, and to avoid solid washout at the top of the digester, where voidage will be high. Details of the necessary calculation for this matching of velocities are given by Boening and Larsen.[45]

Expanded and fluidized bed reactors require additional energy input, as the pressure drop over the bed increases as the bed expands. These systems have, however, an advantage over the packed bed system in that larger areas per unit volume are available for biomass/waste contact. A fluidized bed may have a specific area of 15 times that of a packed bed;[45] thus, increased rates of waste treatment should be possible.

Treating reconstituted sweet whey powder in an expanded bed, Switzenbaum and Danskin obtained high treatment efficiencies at hydraulic retention times of 15 hr and with a waste containing COD of 10 kg/m^3 or below.[46] Under these conditions, soluble COD removals varied between 88 and 91%, although the reduction in total COD was very low. This was due to the large amount of biological solids washed out of the digester, which was sufficient to increase the total COD when treating 5 kg of COD per cubic meter of waste. The solids particles used were aluminium oxide of approximately 0.5-mm diameter.

A similar reactor design was used on undiluted lactic casein whey permeate.[37] Using granular activated carbon particles of 0.5 mm equivalent diameter, 98% of the BOD_5 and soluble COD could be removed at loading rates up to 12.4 kg of COD per cubic meter per day, giving hydraulic retention times as low as 3.8 days. The reduction of both BOD_5 and COD increased with decreasing loading rates, reaching 99.6 and 98.3%, respectively, at 3.8 kg/m³/day. At the high loading rate the gas production rate was 7.7 m³/m³/day, and the methane content in the gas was 52%. At the same conditions, the volatile fatty acid concentrations in the effluent were low, with a total acid concentration of only 135 mg/ℓ. The volatile solids (VS) concentration in the effluent was 6.4 kg/m³.

Dilute reconstituted sweet whey has also been successfully treated in a fluidized bed reactor.[47] At a total COD value in the influent waste of between 6.8 and 7.8 kg/m³, total COD removals of 63 to 68% were achieved at loading rates of 8.5 to 12.2 kg COD per cubic meter per day. Increase in the influent COD to 27.35 kg/m³ saw an increase in the total COD removed to 82%, although the effluent COD nearly doubled.

The fluidized bed reactors have also been investigated for the treatment of dilute lactic casein whey permeate.[45,48] With a waste containing 3.0 kg of COD per cubic meter, 93% soluble COD could be removed at a hydraulic retention time of 9.4 hr. Gas production was 1.9 m³/m³ of fluidized bed volume and day, with a methane content of 65%. This removal efficiency decreased both with increasing waste strength and with decreasing retention times. The digester temperature was 35°C. The biomass in the digester was found to increase with loading rate at this temperature. At a loading rate of approximately 10 kg/m³/day the biomass concentration was approximately 20 kg VSS per cubic meter. This increased to just below 30 kg VSS per cubic meter at 26 kg of COD per cubic meter per day.

A lowering of the digester temperature saw a decrease in the removal efficiencies despite an increase in the biomass concentration in the digester.

VI. ETHANOL FERMENTATION

Ethanol fermentation is one of mankind's oldest technologies. It's history is lost in antiquity, but it is understood that it was known to the Sumerians and Babylonians as early as 6000 B.C. It was not, however, until Pasteur began his work in the 1850s that the microscopic nature of this technology began to be understood. Today the technology is of considerable economical importance both directly and indirectly through taxes, to many nations in the form of the beer, wine, and spirit industry.

Before the advent of the petrochemical industry, all industrial ethanol was produced by the fermentation of carbohydrate substances. Production declined, however, during the 1930s, and has, with few exceptions, disappeared until recently. Ethanol is used today as an intermediate in many chemical synthesis reactions. It is used to produce fat, waxes, olefins, and ethers. It may be used to form detergents, emulsifiers, plasticizers, lubricants, and foaming agents. It is an important raw material for the production of ethylene, itself an important raw material for many petrochemical products. Ethanol is second only to water in its importance as a solvent. It is used as such for drugs, perfumes, cosmetics, lacquers, detergents, and plastics. Recent applications have been in the field of aerosols and mouthwash products. The recent energy crisis renewed the interest in its use as motor fuel, a use which was intensively researched after World War I. Its use as fuel dwindled, however, with the advent of the petrochemical industry, although ethanol-petrol blends were the sole motor spirit sold in France between 1922 and 1935, and were used continuously between 1929 and 1956 in the Mackay district of Queensland, Australia.[49] The recent and continuing uncertainty of the future of the petrochemical industry has led to renewed interest in this process not only for gasohol, but also for industrial solvent production.

A. Biochemistry

The main substrates for ethanol fermentation are the simple carbohydrates of hexoses (glucose, fructose, and galactose), the disaccharides (sucrose, maltose, and lactose), and the trisaccharide maltotriose. The more complex carbohydrates need to be hydrolyzed prior to fermentation, whereas the five-carbon aldopentoses require a special strain of organisms for fermentation.

The conversion of sugars to ethanol by yeast and bacteria proceeds through a series of enzymatic reactions called the Embden-Meyerhor-Parnas (EMP) glycolytic pathway. Under aerobic conditions this pathway provides pyruvate for entry into the tricarboxylic acid cycle, and then ATP, water, and carbon dioxide are produced by oxidative phosphorylation. This gives the maximum available energy for synthesis of cell growth. Under anaerobic condtions the yeast converts the pyruvate to ethanol and carbon dioxide. Less energy is available for cell synthesis under this condition as electron pairs are available for oxidation in the ethanol.

The conversion of lactose to ethanol is essentially the same as for glucose, with the exception of lactose hydrolysis and transport. The lactose is transported across the yeast wall by an inducible and specific enzyme system. Inside the cell it is hydrolyzed to glucose and galactose by the enzyme β-galactosidase. The glucose is converted to ethanol via the EMP pathway, whereas galactose is converted first into the EMP intermediate of D-glucose-6-phosphate by a three-enzyme reaction step before being converted to ethanol.[50]

B. Microbiology

There are a large number of organisms which are capable of fermenting simple sugars to ethanol. Strains from the *Saccharomyces* genus are the most widely used; these strains are, however, unable to ferment lactose. The situation is the same for the bacterial strain *Zymomonas mobilis,* which has received considerable attention for its high ethanol yield and rate of production. Early studies on lactose fermentation indicated that strains of *Torula cremoris,* subsequently renamed *Candida pseudotropicalis,* were the most suitable stains. One study indicated that 46 g/ℓ lactose could be utilized in 22 hr, with an ethanol yield of 80% of the theoretical maximum.[51] Another study gave a somewhat slower fermentation, but a selected strain of *C. pseudotropicalis* completely utilized the 50 g/ℓ lactose in the whey.[52] All other strains utilized less than 76% of the lactose in the same period (55 hr). Among the yeasts tested were strains of *Saccharomyces fragilis,* subsequently renamed *Klyveremyces fragilis* and then more recently *K. marxianus.* Recent investigators have favored the use of yeast of the gender *Klyveromyces.*[50,53-59] The reason for this preference is not often stated, as *C. pseudotropicalis* would appear to be faster,[60] with as good an ethanol yield as any other strain. This preference may, however, be due to the very slight pathogenic risk which is associated with strains of *C. pseudotropicalis,* or the fact that these strains do not form ascospores, and therefore are less easily manipulated genetically.

1. Concentrated Whey

When fermenting concentrated whey there appears to be some controversy as to the best yeast strain(s). When fermenting 100 g/ℓ lactose in whey, the two strains of *K. marxianus* Y18 and Y42 were found to be the greatest ethanol producers when compared with 26 other strains.[50] A similar result was obtained by Moulin et al.[53] and Burgess and Kelly.[60] The former investigators found that *K. marxianus* CBS 397 gave a better performance than *C. pseudotropicalis* IP 513, whereas the latter found that *K. marxianus* NRRL Y1109 utilized lactose at a considerably faster rate than *K. marxianus* (CBS 5795) and *C. pseudotropicalis* (NCYC 744). At 150 g/ℓ lactose all studies found that the appropriate strain of *C. pseudotropicalis* gave the best performance, whereas at higher lactose concentrations there was either no significant difference,[52] or the strains of *K. marxianus* gave the best performance.[50] Under complete anaerobic conditions and using a lactose concentration between 70 and 280

g/ℓ, the strain *C. pseudotropicalis* NCYC 8619 gave the best results when compared with eight other strains.[61]

C. Nutritional Requirements

The majority of investigations into ethanol from whey have been performed without the addition of nutrients, although Burgess and Kelly added yeast extract at a level of 1.0 g/ℓ and urea at 0.5 g/ℓ,[60] and Janssens et al.[54] added 5.0 g/ℓ peptone. It should be noted, however, that both worked mainly with concentrated whey. The effect of adding yeast extract, ammonium sulfate, and potassium phosphate was investigated by Boontanjai.[55] Using fractional factorial design techniques, and investigating the effect of adding yeast extract at the levels of 0.1 and 0.2 g/ℓ, K_2HPO_4 at 0.5 and 1.0 g/ℓ, and $(NH_4)_2SO_4$ at 0.5 and 1.0 g/ℓ, he found that their addition had no significant effect when compared with whey permeate. The parameters tested were lactose and ethanol concentrations after 30 hr and the average rate of lactose utilization as seen by the difference in lactose concentrations at 0, 7.5, 19, and 30 hr into a batch fermentation.

D. Temperature

The optimum temperature for lactose utilization by *C. pseudotropicalis* has been reported to be 37°C.[52] The authors did, however, state that after the yeast had been exposed to 37°C for periods longer than 100 hr, there was little if any difference in the performance of the yeast between 32 and 37°C. The majority of the fermentations reported have also been performed in the range of 30 to 35°C. There appears to be little if any difference when fermenting concentrated whey, although Yoo found that 28°C gave a higher final concentration of ethanol than 32°C but the latter gave a faster production rate when using 200 g/ℓ lactose.[50]

E. pH

PH values between 4 and 6 appear to have little effect on the production of ethanol by either *C. pseudotropicalis* or *K. marxianus*. The optimum pH value has been reported to be between 4.5 and 5.0.[52]

F. Batch Fermentations

There is a large variation in the time required for satisfactory fermentation for ethanol production from whey. Fermentation times vary from 12 hr when using *C. pseudotropicalis* (NCYC 744) in aerated shake flasks with 50 g/ℓ lactose in the whey, to 72 hr for the same fermentation with *K. marxianus* (NCYC 151) and *C. pseudotropicalis* (ATCC 8619) under anaerobic conditions and whey containing 51 g/ℓ lactose.[61] Other workers have reported fermentation times of 16, 17, 22, and 55 hr. These large variations may be due to the differences between aerobic and anaerobic fermentations, as the fermentation time of 72 hr was obtained under anaerobic conditions, whereas the other fermentations were conducted under mildly aerobic conditions. However, it is possible that they are due to the real differences between the strains used as similar differences have been observed elsewhere. Table 5 summarizes the sugar utilization, ethanol productivity, and yield for different reported fermentations.

1. Concentrated Whey

The use of concentrated whey as a raw material would be beneficial as it would lead to an increase in ethanol concentration, thus significantly decreasing the product recovery cost, one of the more important costs besides that of the raw material.[62] Concentrated whey may also be stored during the dairy off-season, thus extending an otherwise seasonal production into a year-long one.

Table 5
SUMMARY OF FERMENTATION PERFORMANCES
FOR DIFFERENT BATCH FERMENTATIONS

Time	Initial sugar conc. (g/ℓ)	Ethanol conc. (g/ℓ)	Ethanol yield (%)	Strain[a]	Ref.
10	50	nq	79	Cp NCYC 744	60
12	50	25	90	Cp Ip 513	53
12	50	25	90	Km CPS 397	53
12	50	nq	90	Km NRRL 1109	60
16	40	18	94	Km Y42	55
16	50	28	87	Km CBS 5795	60
22	46	nq	80	Cp ATC 2512	51
55	50	nq	91	Cp #2	52
72	51	26	97	Cp NCYC 143	61[b]
72	51	22	82	Cp ATTC 8619	61[b]
72	51	21	77	Cp NCYC 744	61[b]

[a] Cp = *Candida pseudotropicalis*; Km = *Klyveromyces marxianus*.
[b] Anaerobic conditions.

Table 6
PERCENTAGE ETHANOL YIELD BASED ON
THEORETICAL MAXIMUM AND INITIAL LACTOSE
CONCENTRATION FOR CONCENTRATED WHEY
FERMENTATIONS

Initial sugar concentration (g/ℓ)						
100	150	200	250	300	Strain[a]	Ref.
90	87	86	68	50	Km CBS 397	53
87	89	86	68	55	Cp IP 513	53
91	53	30	12	—	Km NRRL 1109	66[b]
88	84[c]	—	—	—	Km NRRL 1109	60
93	92	—	—	—	Cp IBS 5795	60
83	86	—	—	—	Cp NCYC 744	60
>100	>100	>100	30	—	Cp ATTC 8619	61[d]

[a] Cp = *Candida pseudotropicalis;* Km = *Klyveromyces marxianus.*
[b] All data based on total solids which contains 70 to 80% lactose.
[c] Yield based on 5% residual lactose.
[d] Anaerobic conditions.

Table 6 summarizes the theoretical ethanol yield based on the initial lactose concentration. It can be noted that the majority of the strains tested were capable of giving a good yield when the whey was concentrated up to 150 g/ℓ lactose. At higher lactose concentrations the efficiencies decreased rapidly for most strains due to incomplete sugar utilization. From the product recovery point of view, this causes few if any problems, as there is no reduction in the recovery cost when the ethanol concentration exceeds 9% (approximately 180 g/ℓ sugar).[62]

The time required for the fermentation increases with increasing sugar concentration. At 100 g/ℓ of lactose the time was found to be between 20 and 22 hr for *C. pseudotropicalis* NCYC 744 and between 24 and 30 hr for *K. marxianus* CBS 5795, the exact times depending on the fermentation temperature. At 150 g/ℓ lactose the times were approximately 30 and 35 hr for MCYC 744 and CBS 5795, respectively.[69] Under anaerobic conditions the fermentation times were considerably longer, with the fastest strain *C. pseudotropicalis* ATTC 8619 requiring 144 hr at 100 g/ℓ lactose for complete fermentation. At 150 g/ℓ lactose approximately 10 g/ℓ lactose remained even after 216 hr. It should, however, be noted that the ethanol yields were higher under anaerobic conditions than under aerobic ones.[61]

G. Continuous Stirred Tank Fermentation Systems

Although the continuous operation of fermentation plants producing ethanol from whey have been investigated and installed, there is a demonstrative lack of data for such systems. A large-scale plant has been installed at Tirau in New Zealand which ferments 1500 m³ of lactic casein whey daily. The plant utilizes deproteinated whey, heat precipitated, and a fermentation system consisting of three fermenters in series, and operates with cell recycle. The plant produces 32,000 ℓ of 95% ethanol daily.[63] A similar system has been developed in Denmark by the Danish Fermentation Industries.[64] This system utilizes, however, only two stages.

A single-stage continuous system has been tested on concentrated whey. Using an ultrafiltration system to achieve 100% recycle of the yeast *K. marxianus* (CBS 397), the fermentation time for 100 g/ℓ lactose could be reduced to 6.7 hr. At this time, lactose utilization was reported to be 100% and the ethanol concentration 47.3 g/ℓ. This gave a theoretical ethanol yield of 88%. Only minor improvements could be made in the ethanol yield and concentration, as it required 20 hr to achieve 96% and 52 g/ℓ for yield and concentration, respectively.[54]

At 120 g/ℓ lactose the fermentation time had to be increased to somewhere between 10 and 20 hr before 100% utilization of the sugar was achieved. At higher sugar concentrations practical operating conditions were not achieved: at 150 g/ℓ lactose the residual sugar became too high and at 200 g/ℓ steady-state conditions could not be obtained.

H. Plug Flow Systems

1. Cell Immobilization

Immobilization techniques have been applied to ethanol production from whey with a view to decreasing the fermentation time. A strain of *K. marxianus* (NRRL Y1109) immobilized in polyacrylamid gel-fermented deproteinated lactic casein whey in 7.7 hr.[65] The conversion efficiency for this process was 79%, and decreased rapidly as the fermentation time decreased.

Using ultrafiltration permeate as the fermentation medium, a strain of *K. marxianus* (B-1-5) maintained stable operation over a period of 10 days when immobilized in calcium alginate.[56] During this period the ethanol concentration was maintained above 20 g/ℓ from 50 g/ℓ lactose, but this dropped off significantly after 10 days of continuous operation. If electrolytically demineralized whey was used instead, the stability could be extended to over 30 days. The lactose utilization varied, however, between 80 and 90%.

2. Tower Fermentations

Utilizing flocculant strains of *K. marxianus* (Y 42), Boontanjai investigated the use of tower fermenters for whey-ethanol fermentation.[55] Full details of the work are reported elsewhere, but it can be noted that fermentation times as low as 1.5 hr were obtained with an ultrafiltrate from lactic casein whey containing between 40 and 50 g/ℓ lactose. The ethanol yield varied between 80 and 97% of the theoretical maximum based on the lactose utilized, with a utilization rate of between 90 and 97%.

Such a fermentation system has considerable advantages over immobilized cell systems, the main ones being that no extra cost is incurred in the form of immobilization chemicals, and the total fermenter volume is available for the fermentation.

I. Use of Prehydrolyzed Whey

In an effort to overcome the problems of so few organisms being capable of utilizing lactose as a fermentation medium, O'Leary and co-workers investigated the use of prehydrolyzed whey, thus making the feremntation substrate a mixture of glucose, galactose, and any residual lactose present. Using the strains *K. marxianus* (NRRL 1109) and *Saccharomyces cerevisiae* (ATTC 834), their initial work was performed using cottage cheese whey where hydrolysis was carried out by β-galactosidase prior to coagulation.[57] The results showed clearly that there was no benefit in prehydrolyzing the whey. Although the growth rate and the final cell concentration of *K. marxianus* were higher with the hydrolyzed whey, the rate of production of ethanol was considerably slower. The maximum ethanol concentration was achieved at 120 hr for the hydolyzed whey, whereas the control required only 72 hr. The reason for this slow production became clear when examining the concentration profiles of the individual sugars. In the first 24-hr period, glucose disappeared at approximately the same rate as lactose in the control. Despite this apparent equality in sugar utilization rates in this period, the hydrolyzed whey produced ethanol at a rate faster than that of the control. In this period, however, there was very little galactose utilization, and this sugar was not utilized before the glucose was consumed. The rate of galactose consumption was found to be considerably slower than the rate of lactose consumption, with a similar slow rate of ethanol production.

When utilizing *S. cerevisiae* a similar pattern occurred. Glucose was utilized within the first 24 hr, but the lag time experienced for *S. cerevisiae* at 5.76 hr was much longer than that for *K. marxianus* (2.73 hr) under the same conditions. If the yeast had been pregrown on glucose, it did not grow or produce ethanol from the galactose fraction; thus, the maximum ethanol concentration (8.5 g/ℓ) was obtained after 24 hr. If the yeast was pregrown on galactose, it did grow and produce ethanol on the galactose fraction, but it required between 48 and 96 hr to adjust from glucose to galactose before such a growth was restarted. The maximum ethanol concentration was reached at 120 hr, although there was approximately 10 g/ℓ galactose remaining at this point. After this there was a concurrent decrease in galactose and ethanol concentration until the termination of the fermentation at approximately 210 hr.

Subsequent work on concentrated hydrolyzed whey gave similar results in that, although *S. cerevisiae* was a more efficient glucose fermenter than *K. marxianus*, neither organism appeared to perform significantly better than *K. marxianus* on nonhydrolyzed whey.[66] *S. cerevisiae* utilized 100% of the glucose up to a TS concentration in the hydrolyzed whey of 300 g/ℓ, whereas *K. marxianus* utilized 100% of the glucose up to 200 g/ℓ of TS. Utilization of glactose occurred, however, only with *K. marxianus* at a TS concentration of 50 and 100 g/ℓ, where the utilization rates were 100 and 30%, respectively. *S. cerevisiae* did not utilize galactose at all, thus making this route to ethanol from whey totally uneconomical as galactose contributes 50% of the total sugar concentration.

J. Effect of Substrate and Product Concentrations

As previously described, the rate and efficiency of ethanol fermentation decreased as the lactose concentrations increased. Furthermore, when high lactose concentrations were used, the fermentations stopped before the lactose was fully utilized. The cause of this is inhibition by ethanol. In their study, using yeast nitrogen base and added lactose, Moulin et al.[67] found that ethanol was inhibitory to *C. pseudotropicalis* IP 513, even when present in concentrations as low as 16 g/ℓ. The inhibition increased markedly as the ethanol concentration increased. Furthermore, they found that there were no evidence of inhibition on the fermentation by lactose up to 250 g/ℓ when no ethanol was present. There were, however, signs of substrate inhibition when the ethanol exceeded 64 g/ℓ and lactose 175 g/ℓ. They also stated that ethanol was more inhibitory to this strain when in the presence of lactose than when in the presence of glucose. This may explain the observed effect by O'Leary et al.[66] in that the solids in hydrolyzed whey permeate were less inhibitory than the solids in whey permeate at the same concentrations.

1. Strain Improvements

Improvements in the tolerance of the organism to ethanol can be achieved either by adaptation or by the addition of lipids to the medium. Boontanjai showed that the strain *K. marxianus* UCD FST 7158 was incapable of utilizing more than 68% of the lactose in concentrated whey permeate with 100 g/ℓ lactose.[55] After a series of subculturing, exposing the organism to increasing concentrations of ethanol, it utilized 90% of the lactose within 24 hours after only four such subculturings. The addition of oleic and linoleic acids and ergosterol to the fermentation medium was shown to decrease the fermentation time for *C. pseudotropicalis* CBS 397 from 90 to 60 hr when fermenting concentrated whey.[68]

VII. YEAST PRODUCTION

An excellent review on the production of biomass from whey has recently been published by Meyrath and Bayer.[69] It is therefore not the intention of the authors to cover this topic in any detail here. The research and production of microbial protein received considerable interest during the so-called food crisis of the 1960s. Several large-scale plants were proposed using a variety of raw materials including *n*-paraffins, methanol, and methane. The vast majority of these plants never came into production, either through consumer resistance to their product or to the advent of the energy crisis in the 1970s. It is, however, the author's understanding that the ICI plant at Teeside is still in production, using a methanol-oxidizing bacterial culture.

The use of yeast for human consumption is a well-established practice,[69] particularly if the yeast is a *Saccharomyces* species, and the practice of using yeast extract as a sandwich spread is well known to all youngster in Australasia. It has also been shown that mixtures of yeast and whey proteins are superior to soymeal as a protein source for piglets.[70]

In the production of yeast biomass from whey, supplementation of nitrogen is required, as whey proteins are nonassimilable to yeasts. Supplementation of minerals other than sulfurs is usually not necessary, as the current theory is that there are sufficient minerals in whey. This may not, however, be an accurate statement, since interaction may occur.[69] It has also been shown for other fermentations that higher-order interactions between medium components are of significant importance.[71] Addition of vitamins has also been found to enhance the feremntation, whereas the addition of phosphate and yeast has been questioned if ammonium hydroxide is used as a neutralizing agent and magnesium sulfite or calcium chloride is added.

Meyrath and Bayer list three modern yeast production processes from whey.[69] The Bel Formangeries process uses deproteinized whey diluted to a lactose concentration of 3.4%. The fermentation process is continuous with a fermentation time of 4 hr, and it operates at a temperature of 38°C and a pH of 3.5. The organism is *K. marxianus,* and the aeration rate is approximately 1.2 volumes of air per volume of fermenter per minute.

The Kiel process uses two organisms. *Lacobacillus bulgaricus* converts the lactose to lactic acid, which is then utilized by *Candida krusei.* The fermentation time in the first stage is between 7 and 11 hr and is 7 hr in the second stage. Both fermenters operate above 40°C.

The Vienna process uses an acid-resistant strain of *C. intermedia.* No pretreatment of the sweet whey is necessary, but ammonia and ammonium sulfite is added to the whey. The productivity of the process is 4.5 kg/m³/hr, with a final cell concentration of 24 kg/m³.

VIII. MICROBIAL OILS AND FATS

Certain microorganisms, including yeasts, filamentous fungi, and algae, can accumulate considerable quantities of lipids when grown under appropriate environmental conditions. The industrial potential of this has been recognized for many years and some excellent reviews have appeared on the subject.[72,73] Although interest in microbial fat declined considerably after World War II, it has recently been rekindled both because of the need to utilize industrial wastes and because of the technology development for large-scale continuous fermentation processes, particularly for single-cell protein production. The possible uses of microbial fats and oils are similar to those from plant and animal sources although, of course, much depends on the composition of the microbial lipid. This is predominantly triglyceride with minor amounts of sterols and hydrocarbons. The most commonly encountered fatty acids are palmitic, stearic, linoleic, and linolenic.[74] Production of microbial oil is typically favored using a substrate which is limited in one nutrient, often nitrogen. Compared with the production of single-cell protein, the requirement for aeration is relatively low. In a typical fermentation process, fat accumulation follows a two-stage pattern.[73] Initially, the organism grows at its maximum rate until a nutrient other than carbon, usually nitrogen, becomes exhausted. Growth then ceases but the residual carbon substrate continues to be consumed and is converted into lipid. Fat contents of up to 80% of the dry biomass can be achieved depending on the organism and fermentation conditions.[73] Given the need for a nitrogen-deficient substrate, whey (permeate) may prove to be suitable provided that appropriate microorganisms can be recognized.

Wix and Woodbine examined 40 species of fungi for their ability to accumulate fat when growing on whey.[75,76] The most promising organisms in terms of lactose consumption and biomass produced were *Aspergillus ustus*, *Penicillium frequentans*, *P. oxalicum*, and *P. notatum*. *A. ustus* used up to 96% of the lactose in the whey and produced biomass of 17 g/ℓ containing 30% fat. Mickle et al.[77] adapted the yeast *Rhodotorula gracilis* to grow on whey but lactose utilization was poor. Only when the whey was enriched with sucrose was a lipid content of 53% of the biomass obtained.

Cheese whey and whey permeate have been used as substrates for oil production using two strains each of the yeasts *Candida curvat* and *Trichosporon cutaneum*.[78,79] The strains were originally isolated from cheese factory floors and drains. Fermentation of whole whey by these yeasts was relatively inefficient and produced low yields of oil. However, fermentation of the permeate was much more successful. *C. curvata* D produced the most oil, reduced the COD of the permeate by 95%, and required the fewest additions of supplementary nutrients (0.3 mℓ/ℓ permeate of 15.5% NH_3). A biomass production of 27 g/ℓ, containing 57% oil, was obtained after 72 hr of fermentation. The fat contained oleic (50%), palmitic (30%), stearic (15%), and linoleic (8%) acids. The cell protein was rich in methionine and other amino acids except lysine, and could, therefore, be used as an animal feedstuff after extraction of the oil. Different methods of extraction were investigated, and the best results were obtained by sequential extraction using methanol and benzene or ethanol and hexane.[78]

As mentioned above, a nitrogen-limited substrate is conducive to oil accumulation by appropriate microorganisms. Hammond et al.[74] varied the nitrogen to carbon ratio of cheese whey permeate to determine its effect on oil production by *C. curvata* D. When the ratio was increased relative to the normal situation, the rate of COD reduction was increased but the amount of oil produced per dry biomass was decreased. In contrast, when the ratio was decreased, the fermentation time was increased and the total biomass produced was decreased. However, the amount of oil produced per dry biomass was greatly increased. During studies in continuous culture using *C. curvata* D (subsequently reclassified as *Apiotrichum curvatum* ATCC 20509) grown on whey permeate, Davies and Gordon obtained maximum oil yields at a total nitrogen concentration in the feed permeate of 465 mg/ℓ (obtained by addition of ammonia to the permeate).[80]

Since fermentation processes require considerable energy for medium sterilization, any process which can operate successfully under nonsterile conditions will be economically advantageous. Hammond et al.[74] demonstrated that ultrafiltration of whey typically removes greater than 99% of the microbial load, and the fermentation process can subsequently be successfully performed in a nonsterile, but sanitized, fermenter (rinsed with a 200-ppm chlorine solution), provided that the yeast inoculum is 10^6 cells per milliliter or greater. If the permeate is pasteurized (63°C, 30 min), a minimum inoculum of 10^5 cells per milliliter is required. During their studies in continuous culture, Davies and Gordon identified an optimum pH value for oil production of pH 3.5,[80] a value which makes nonsterile operation on an industrial scale feasible.

Two recent studies have been described using *C. curvata* D grown in continuous culture on whey permeate.[80,81] Both studies identified a low dilution rate as being optimal for oil production. At a dilution rate of 0.05^{-1}, Floetenmeyer et al.[81] reported a COD reduction of the permeate of 92%, and a biomass production of 13.3 g/ℓ containing 46% oil. This represents a productivity of 0.31 g oil per liter per hour. Davies and Gordon,[80] using a dilution rate of 0.4^{-1} and a total nitrogen concentration in the feed medium of 465 mg/ℓ, obtained a biomass concentration of 20 g/kg culture, containing 46% oil and representing a productivity of 0.37 g oil per kilogram of culture per hour. At higher dilution rates residual lactose appeared in the effluent. It was possible to achieve full lactose utilization at higher dilution rates if the nitrogen concentration of the permeate was increased. Unfortunately, however, this led to lower yields of oil.[80] It was also commented that the fatty acid composition of the lipid remained relatively constant at different dilution rates and nitrogen contents of the permeate. The composition described was similar to that detailed above, and the oil was reported as being similar in physical and chemical characteristics to palm olein, a product which is widely used as a bulk liquid frying oil.[80]

Davies and Gordon have presented a flow scheme for a process plant with a whey feed intake of 20 m³/hr operating 300 days/year (typical for a New Zealand dairy factory).[80] Given a value of N.Z. $800 per tonne for the yeast oil, the internal rate of return was estimated to be greater than 30%. It appears, therefore, that provided the nature and properties of the microbial lipid are such that a ready market can be found, whey disposal by this process may be profitable. Further, it may be possible to manipulate the fatty acid composition of the lipid to meet any requirements by the application of mutation or genetic recombination techniques.

IX. BUTANOL

Production of *n*-butanol by fermentation was a well-established commercial process prior to World War II, reflecting the importance of this product as a chemical feedstock. Since then, however, cheaper petrochemical sources have replaced the fermentation route, although interest has revived over the last few years. This revival has been due, at least in part, to the potential use of butanol as a fuel supplement. As a gasoline extenter, butanol is superior to ethanol because of its greater energy per unit mass. Further, butanol can be mixed directly with diesel fuels for use in compression ignition engines, and it can also be used as a cosolvent to prevent phase separation in methanol/gasoline blends.[82] The fermentation process is traditionally known as the acetone-butanol-ethanol (ABE) fermentation since all three chemicals are produced during the process. Butanol is the predominant product with ratios ranging from 3:6:1 to 1:10:1 (A:B:E) being reported.[83] Details of the traditional process have been well documented,[83] and Spivey has recently described the operation of the process at National Chemical Products, South Africa.[84] Typically, the substrate used has been molasses, and the fermentation is carried out using various strains of the anaerobic bacteria *Clostridium acetobutylicum* or *C. beyerinckii*. The sugar concentration is normally restricted to 50 to 60 g/ℓ as the process suffers from severe product inhibition at total solvent concentrations of approximately 16 g/ℓ.[83,85] However, it is this very fact that renders whey a suitable fermentation substrate since the lactose concentration rarely exceeds 50 g/ℓ.

Maddox described the use of sulfuric acid casein whey permeate as a substrate for butanol production using *C. acetobutylicum* NCIB 2951.[86] When the permeate was supplemented with yeast extract (5 g/ℓ) a butanol concentration of 15 g/ℓ was obtained after 5-days fermentation, representing a yield of 0.36 g/g lactose utilized. Ethanol and acetone concentrations approximated 1.5 g/ℓ. The progress of the fermentation was similar to that using a molasses substrate, although the fermentation time was longer. Initially, lactose was utilized for cell growth accompanied by production of acetic and butyric acids (acidogenic phase). Subsequently, the acids and the residual lactose were utilized for butanol production (solventogenic phase). When using whey permeate which had not been supplemented with yeast extract the fermentation was slower, and the maximum butanol concentration of 13 g/ℓ was not reached until after 7-days fermentation. Welsh and Veliky performed similar experiments using *C. acetobutylicum* ATCC 824 grown on acid whey.[87] A total solvent concentration of 9.2 g/ℓ was obtained after 5-days fermentation at 37°C. The use of nonsterilized whey and/ or agitation during the fermentation was detrimental to solvent production. The reason why agitation should have this effect is not entirely clear, although it may be related to the suggestion that the maintenance of a positive head-space pressure during the fermentation is beneficial to butanol production.[88] Thus, agitation may assist stripping of hydrogen gas (a by-product of the process) from the medium, and thus cause a loss of reducing power to convert butyric acid to butanol.

Schoutens et al.[89] used *C. beyerinckii,* which produces only two solvents, butanol and isopropanol, to investigate production from whey ultrafiltrate, glucose, lactose, and galactose. Production from ultrafiltrate was enhanced by fermenting at 30 rather than 37°C. Compared to a glucose substrate, butanol production from whey ultrafiltrate was lower, although the overall product yield, based on sugar utilized and productivity, were comparable. Typical results using ultrafiltrate at a TS concentration of 50 g/ℓ, supplemented with yeast extract, were a butanol concentration of 5 g/ℓ (isopropanol 0.9 g/ℓ), representing a yield of 0.24 g/g lactose utilized, and a mean reactor productivity of 0.4 g/ℓ/hr. Productivities obtained by other workers include 0.125,[86] 0.076,[87] and 0.180 g/ℓ/hr.[88] To assess the feasibility of using hydrolyzed whey permeate as the fermentation substrate, galactose and mixtures of glucose and galactose were tested as carbon sources for *C. beyerinckii,*[89] When present on its own, galactose was less readily used than glucose, but product yields based on sugar utilized were comparable. When present in mixtures, glucose was used in preference to galactose.

All of the studies described above were performed using laboratory-scale, e.g., 2-ℓ, fermenters. Experiments have recently been undertaken jointly by the New Zealand Dairy Research Institute and the Biotechnology Department, Massay University, New Zealand, to investigate butanol production on the 100-ℓ scale.[90] Substrates used were sulfuric acid casein whey permeate and serum derived from heat-acid precipitation and separation of (lactalbumin) whey protein. The substrate was supplemented with ammonium ions and/or yeast extract, and adjusted to pH 6.5 using either ammonia or, in the case of yeast extract supplementation, sodium hydroxide. The fermentations were conducted at 32°C, without agitation, and with an evolved gas pressure of 65 KPa. Results are described in Table 7. In general, supplementation with yeast extract or ammonia gave similar butanol production, but maximum productivity was achieved when both were present.

Very little information is available with reagrd to the optimum pH of the butanol fermentation when using a whey (permeate) substrate. Maddox adjusted sulfuric acid casein whey permeate to pH 6.5 prior to autoclaving, but the value at inoculation was pH 5.3.[86] This subsequently dropped to pH 4.45 as acids were produced from lactose, but rose to pH

Table 7

**PRODUCTION OF BUTANOL FROM WHEY PERMEATE
AND LACTALBUMIN SERUM BY FERMENTATION
WITH *C. ACETOBUTYLICUM* NCIB 2951**

	Supplement				
	Yeast		Fermentation	Butanol	Lactose
	extract	NH_3	time	conc.	utilization
Substrate	(g/ℓ)	to pH	(hr)	(g/ℓ)	(%)
Permeate	10	—	150	6.5	64
Permeate	—	6.5	150	6.2	46
Serum	3	6.5	150	8.0	46
Serum	10	6.5	111	8.4	64

4.9 by the completion of solvent production. Subsequent experiments were performed where the pH was not allowed to fall below pH 5.0, but this proved detrimental to solvent production.[88] Welsh and Veliky adjusted acid whey to pH 6 prior to autoclaving,[87] and the value at inoculation was pH 5.7. Schoutens et al.[89] compared initial values of pH 5.2 and 6.5 for whey ultrafiltrate, but the solvent production was little affected.

As mentioned above, one of the problems of the fermentation process is that it suffers from product inhibition. This effect has been quantitated using *C. acetobutylicum* NCIB 2951 growing on whey permeate.[85] Each of the three products was tested independently by adding it to the fermentation 24 hr after inoculation, when cell growth was almost completed but solvent production was only 1 g/ℓ. In the presence of 10 g/ℓ of added butanol, production rates were virtually unaffected, but the final butanol production was reduced from 8.6 to 5.8 g/ℓ. At 20 g/ℓ of added butanol, production ceased very soon after solvent addition. Ethanol, at levels of up to 30 g/ℓ, had little effect on solvent production, but at 50 g/ℓ inhibition was observed. Acetone, at 10 g/ℓ, had little effect, but at 20 g/ℓ the final solvent production was reduced.

Another problem which may occur with this fermentation process is culture degeneration. This is manifest by cultures, after repeated subculturing, being unable to produce solvents. Such an effect has been described by Gapes et al.[91] using *C. butylicum* NRRL B-592 growing on whey permeate. When growing optimally a total solvent concentration of 11.2 g/ℓ was achieved, but after repeated subculturing no solvents were produced although lactose was still being utilized. Thus, it is clear that methods of culture maintenance and inoculum development should be such that the number of culture transfers is minimized.

Traditionally, commercial production of butanol has been conducted using conventional batch fermentation techniques, and the majority of the research studies using whey have also employed this technology. However, continuous fermentation techniques would be economically beneficial and would provide higher productivities.

Krouwel et al.,[92] using a glucose substrate, have demonstrated the use of calcium alginate-immobilized cells for continuous production of butanol and isopropanol. Productivities up to 0.8 g/ℓ/hr were reported, with the maximum duration of an experiment of 290 hr.[93] Monot and Engasser used free cells of *C. acetobutylicum* ATCC 824 for continuous fermentation of glucose and achieved a solvent productivity of 0.4 g/ℓ/hr. Very recently, Schoutens et al.[94] have achieved the continuous production of solvents from whey permeate using calcium alginate-immobilized *C. beyerinckii*. They found that both a fermentation temperature of 30°C and a dilution rate of 0.1 hr^{-1} or less during the start-up phase were required to obtain continuous production. At an operating dilution rate of 0.7 hr^{-1} a maximum solvent productivity of 1.0 g/ℓ/hr was observed, a value considerably higher than any obtained in batch fermentation.

The studies described above have clearly demonstrated the feasibility of butanol production from whey, although lactose is a less preferred substrate than glucose or sucrose. It is anticipated, however, that this disadvantage can be minimized with the application of new fermentation technology and the isolation of new bacterial strains with a greater affinity for a lactose substrate.

Several economic evaluations have been reported for a commercial process producing butanol from whey.[82,88,95,96] Lenz and Moreira examined a plant with a production capacity of 45 · 10[6] kg solvents per year, processing between 10,000 and 12,000 m[3] of whey per day, and estimated a net return on investment of 31%.[95] However, the costs were based on the assumption that whey would be as satisfactory a substrate as molasses. Volesky et al.[96] made the same assumption and estimated a butanol production cost of U.S. $0.29 per kilogram. Gapes used his own experimental data from 10-ℓ scale fermentations using a substrate of sulfuric acid whey permeate, supplemented with yeast extract, to design and cost a plant operating under typical New Zealand conditions.[82] Taking a whey flow rate of 300 m[3]/day for 270 days/year, and a solvent production of 14.4 g/ℓ after 65 hr of fermentation, the average solvent cost was N.Z. $1.57 per kilogram. For a 330-day year the cost reduced to N.Z. $1.23 per kilogram. If the throughput were doubled the costs would be N.Z. $1.28 and N.Z. $0.96 per kilogram for 270 and 330 days of operation, respectively. It was concluded that the latter situation could be economic, particularly if fermenter costs could be reduced, e.g., by the introduction of novel fermentation technology. In this regard, Larsen has studied the effects on cost of both reducing the fermentation time and increasing the total solvent concentration in the broth.[97] Such improvements could be made by the application of novel fermentation technologies and/or isolation of improved strains of organisms. Taking a base level of 44.5-hr fermentation time producing 20 g/ℓ solvents at a cost of N.Z. $0.75 per kilogram, a reduction in fermentation time to 11 hr would reduce the cost by 19%, and to 5 hr by 22%. For an increase in solvent concentration to 25 g/ℓ, a cost reduction of 15% would be achieved, and to 30 g/ℓ the reduction would be 27%. Thus, these calculations allow identification of those aspects of the process where improvements in the economics may be made.

X. BUTANEDIOL

2,3-Butanediol is produced as a fermentation end-product by several bacteria, and has potential as a liquid fuel and chemical feedstock.[98] Dehydration of 2,3-butanediol yields the common industrial solvent methylethylketone, and further dehydration gives 1,3-butadiene, the monomeric subunit in synthetic rubber. Although production by fermentation has never been a commercial process, several pilot-plant operations were developed during World War II.[99] Since then, however, the process has been rather neglected until the last few years. The first report of whey being used as a fermentation substrate was in 1982, when Speckman and Collins used cheese whey to study butanediol production by *Bacillus polymyxa, Streptococcus diacetilactis, Torulopsis colliculosa,* and *S. faecalis.*[100] The most promising results were obtained using *B. polymyxa,* which gave a butanediol concentration of 3.7 g/ℓ after 7 days incubation at 30°C. Strains of this organism have previously been shown to be among the best producers of butanediol.[99] Further experiments showed that although the process is essentially anaerobic, a small amount of aeration is required to achieve maximum production. This is probably because aeration encourages greater biomass production to subsequently convert the sugar to butanediol. Although a butanediol production of 3.7 g/ℓ appears promising, it is in fact rather poor when compared to other substrates.[99] The theoretical yield of butanediol, expressed as grams of product per gram of lactose utilized should approach 40%. In the experiments described, the lactose utilization was 25.2 g/ℓ. On this basis, then, the butanediol yield was only 15%, suggesting that the strain of *B. polymyxa* used may not have been the most appropriate.

Continuous fermentation of whey permeate by *B. polymyxa* has been used to improve the productivity of the process.[101] A stirred tank reactor was used in which the bacterial cells were recirculated into the medium by separation in a hollow fiber ultrafiltration cartridge. A butanediol productivity of 1.04 g/ℓ/hr was achieved compared to 0.024 g/ℓ/hr in batch fermentation. Although the yield, based on lactose utilized, was still only 17%, the addition of acetate ions at a concentration of 75 mM, apparently increased this value to 68%, and the productivity was increased to 8.18 g/ℓ/hr. However, it is not clear from the report whether these increases were due solely to lactose utilization or whether acetate utilization made a contribution.

In an attempt to find a more suitable organism for butanediol production from whey permeate, Lee tested two strains of *Enterobacter aerogenes* and one each of *B. polymyxa* and *Klebsiella pneumoniae*.[102] All of these organisms have been previously reported to be good producers of butanediol.[98,99] *K. pneumoniae* NCIB 8017 proved to be the most promising. Using rennet whey permeate as substrate a butanediol concentration of 7.5 g/ℓ, representing a yield of 46% based on lactose utilized, was achieved after 96 hr of incubation. The overall productivity was 0.08 g/ℓ/hr. Ethanol and acetoin were also produced at concentrations of 0.3 and 1.1 g/ℓ, respectively. Experiments were also reported where hydrolyzed lactic acid casein whey permeate was used as the substrate. In this case, a butanediol concentration of 13.7 g/ℓ was obtained after 96 hr of fermentation (productivity 0.14 g/ℓ/hr). On the basis of these results, therefore, it appears that *K. pneumoniae* is superior to *B. polymyxa* in batch fermentation of whey permeate, and that prior hydrolysis of the lactose is beneficial to the process.

Very recently, alginate-immobilized cells of *K. pneumoniae* have been shown to be useful for butanediol production from whey permeate, and this technique may prove useful in improving the productivity of the process.[103]

So far, little attention has been paid to product recovery of butanediol at the end of the fermentation. Clearly, this area requires some development before any industrial fermentation process can be considered.

XI. LACTIC ACID

Lactic acid is widely used in the food industry although the market is relatively small and growing only slowly.[104] It also has the potential of being a useful chemical feedstock, and calcium lactate is used pharmaceutically as a soruce of calcium. The fermentation process for lactic acid production has been operated commercially for many years, but competes with the synthetic product made from acetaldehyde with hydrogen cyanide. Fermentation lactic acid is generally less expensive than the synthetic product, but usually contains impurities such as residual carbohydrate.[105] Since the postfermentation purification process is relatively expensive, the purer grades of product are generally cheaper to produce by the synthetic route. If, however, the prices for feedstocks used in chemical synthesis escalate, the market for the fermentation product would increase markedly.[106] At present, the application of new fermentation technology and/or improvement of the product recovery process could be used to reduce overall production costs and thus enable high-quality lactic acid to be produced by fermentation at a cost comparable to the synthetic route.

The traditional fermentation process for producing lactic acid from whey has been well described.[107,108] The process is anaerobic but the organisms used are facultative anaerobes and strict exclusion of air is unnecessary. The organisms are all homofermentative species of *Lactobacillus,* usually *L. delbrueckii,* but *L. bulgaricus* is used with whey due to the inability of the former species to use lactose. The fermentation is operated at 45 to 50°C and a slightly acid pH, with control if necessary. Product recovery processes include precipitation of calcium lactate followed by treatment of this with sulfuric acid, precipitation as the more easily purified zinc salt, and purification by solvent extraction or by distillation of the methyl ester.[104]

Marshall and Earle,[109] using batch fermentation with a substrate of lactic casein whey, have demonstrated that lactic acid production occurs in four distinct phases: (1) an "exponential" phase, corresponding to the exponential growth phase of the cells, where the lactic acid concentration increases exponentially with time; (2) a transition period between (1) and (3) when cell growth ceases; (3) a period of constant acid production rate corresponding to the stationary phase of cell growth; and (4) a phase of declining acid production as lactose becomes limiting. Greater than 50% of the total acid production occurred during the stationary phase. The fermentation was conducted at 46°C and pH 6.0 using dried whey powder at a concentration of 70 g/ℓ. The lactic acid concentration after 24 hr of fermentation was 30 g/ℓ, representing an overall productivity of 1.25 g/ℓ/hr. Supplementation of whey with tryptophan, casamino acids, and a number of vitamins is beneficial to lactic acid production, and sodium caseinate is a good source of these nutrients.[110] Although oxygen is inhibitory to cell growth, it has no effect on acid production during the stationary phase.[110] Agitation of the culture is necessary during the process to suspend the fine protein particles.[109]

As a means of reducing production costs, various new fermentation technologies have been applied to the process. Marshall investigated continuous culture, using a substrate of lactic casein whey, and demonstrated that a temperature of 46°C and a pH in the range 5.4 to 6.0 were optimal for acid production by *L. bulgaricus*.[110] Productivities greater than 3 g/ℓ/hr were achieved at a dilution rate of 0.2 hr^{-1}. Despite the use of very low dilution rates (0.3 hr^{-1}) it was not possible to achieve high enough cell concentrations to completely utilize the lactose. It was concluded that although continuous production from whey is feasible, multistage continuous reaction systems and/or cell feedback is necessary to reduce the lactose concentration in the effluent to an acceptable level. Keller and Gerhardt subsequently achieved this goal by employing two fermenters in series.[111] In comparison with a single-stage fermentation, the residual lactose was reduced from 7 g/ℓ to less than 1 g/ℓ. A further refinement of continuous multistage fermentation has been proposed by Setti.[112] In this system the effluent from each fermenter is passed through a reverse osmosis unit where lactic acid, but not lactose, is removed, thus alleviating any problems of product inhibition.

Another menas of countering the problem of product inhibition, and thus achieving higher productivities, is dialysis culture. In this system the product is essentially removed from the culture as soon as possible after its formation. Friedman and Gaden used such a system to confirm the inhibitory effect of lactate on lactic acid production from glucose by *L. delbrueckii*.[113] Since lactate was continuously removed, overall productivities were higher than in conventional batch culture. Subsequently, Stieber and Gerhardt described a dialysis continuous process for ammonium lactate production from cheese whey ultrafiltrate.[114] The fermenter was operated without an effluent, thus effectively immobilizing the cells and achieving high biomass concentrations. The ultrafiltrate (lactose concentration 62 g/ℓ) was supplemented with yeast extract and fed without sterilization into a fermenter maintained at pH 5.5 and 44°C with *L. bulgaricus*. The system was operated without interruption for 26 days. During steady-state conditions the lactate concentration in the dialysis was 35.1 g/ℓ, and an overall productivity of 3.2 g/ℓ/hr was achieved.

Several other novel fermentation technologies have recently been applied to lactic acid production. Unfortunately, however, most authors have used a glucose substrate, but no doubt the principles can be applied to whey. Essentially, these technologies aim to both increase productivity and to ensure no residual sugar in the lactic acid-containing effluent. Calcium alginate-immobilized cells of *L. delbrueckii* have been proved effective in both batch and continuous culture.[115,116] During continuous operation in a packed-bed column reactor the maximum lactic acid yield from glucose was about 83% and the productivity of the system approximated 3 g/ℓ/hr. However, this is no greater than that obtained in conventional continuous culture.[110] The half-life of the immobilized biocatalyst in continuous operation was about 100 days. Immobilized cells of *L. bulgaricus* were shown to be as effective as *L. delbrueckii* using a glucose substrate, but the product yield from lactose was only 40%.[116] Vick Roy et al.[117] applied a hollow fiber fermenter to lactic acid production from glucose by *L. delbrueckii*. The advantage of this technique is that extremely high cell concentrations, and thus high productivities, can be attained. Productivities of up to 100 g/ℓ/hr were reported for the system which are far higher than other reports to date. Unfortunately, however, the glucose conversion was low and it is apparent that some productivity must be sacrificed to increase this conversion. To this end, a system of cell recycle, where the cells are separated from the effluent using an ultrafiltration unit, has been applied to a conventional continuous fermentation process.[105] Using a glucose substrate at 50 g/ℓ a lactic acid productivity of 76 g/ℓ/hr was recorded. Lactic acid and glucose concentrations in the effluent stream were 35 g/ℓ and less than 0.20 g/ℓ, respectively, while the cell concentration in the fermenter was 54 g/ℓ. Thus, the dual aims of high productivity and low residual sugar were realized. A preliminary economic analysis of this technology suggested that for a production rate of 5×10^6 kg/year, up to 10% of the production cost could be saved over conventional technology using batch fermenters.[105]

The future, then, for fermentation lactic acid looks promising. However, the development of new product recovery techniques is still awaited.

XII. CITRIC ACID

Until the early part of this centruy, commercial citric acid was produced mainly from lemon juice. Today it is produced almost exclusively from fungal fermentations, and annual production is estimated at over 300,000 tons.[118] It is used extensively in the food and pharmaceutical industries, where it acts as an acidulant, and has also found application in detergent formulation and the treatment of boiler water. Many microorganisms are known to accumulate citric acid, but only selected strains of the fungus *Aspergillus niger* are used commercially. Recently, yeasts of the genera *Candida* and *Saccharomycopsis* have been investigated, principally because of their ability to use paraffins as substrate.[118] However, the increase in oil prices over the last decade has slowed any commercial development of oil derivatives as a cheap fermentation substrate.

Processes for the production of citric acid by *A. niger* have been the subject of several reviews.[118-121] In the Western world about 80% of the citric acid used is produced by the submerged fermentation process. Substrates typically used are molasses and glucose syrups. The initial sugar concentration is adjusted to about 150 g/ℓ and after about 10 days of fermentation, citric acid yields of up to 90% are achieved. Attempts to use whey or whey permeate as the substrate have been reported only recently. Somkuti and Bencivengo tested five strains of *A. niger* for their ability to produce citric acid in acid whey permeate. All of the strains grew well, indicating that the permeate had an adequate nutritional status.[122] The best results were obtained using *A. niger* NRRL 599, which gave a citric acid concentration of 5.2 g/ℓ after 8 days of fermentation. This represented a yield of 22%, based on lactose utilized, but only 52% sugar utilization occurred (initial concentration 44 g/ℓ). The course of the fermentation was similar to that using other substrates. Initially, mycelial growth occurred with little acid production, but once growth ceased, after about 4 days, citric acid production commenced. After reaching a maximum, however, citrate levels decreased despite the presence of residual lactose. Hossain et al.[123] examined 11 strains of *A. niger* for their ability to produce citric acid from lactic casein whey permeate. Some strains grew poorly, with poor utilization of lactose, but five of the strains produced citric acid. The best results were obtained using *A. niger* IMI 41874, which gave a citric acid concentration of 5.0 g/ℓ, representing a yield of 13.5% based on lactose utilized. The course of the fermentation was similar to that previously described.[122] Interestingly, the Commonwealth Mycological Institute indicates that *A. niger* IMI 41874 is identical to *A. niger* NRRL 599.[124]

The results described above, while showing the feasibility of citric acid production from whey permeate, are relatively poor when compared to other substrates. Of particular concern are the low lactose utilization and the poor yields based on lactose utilized. Furthermore, industrial processes use much higher initial concentrations of sugar (150 g/ℓ), suggesting that if whey permeate is to be a competitive substrate it must be either supplemented with lactose or used in a concentrated form. Both Hossain et al. and Somkuti and Bencivengo isolated mutant strains of their respective organisms and demonstrated these to be superior citric acid producers.[122,123] When the permeate was supplemented with additional lactose to 140 to 150 g/ℓ, improvements were noted in both citric acid production and yield, but the total lactose utilization remained relatively poor. Productions of 28 g/ℓ (yield 57%) and 15 g/ℓ (yield 23%) were obtained by Somkuti and Bencivengo and Hossain et al., respectively.[122,123] The additional presence of methanol (up to 3% v/v) was also stimulatory to production,[122,123] as was supplementation of the permeate with an additional nitrogen source.[123]

During the fermentation of acid whey permeate, the culture pH typically decreases from its initial value of pH 4.5 to pH 1.8. Hossain et al.[123] attempted to improve citric acid production by controlling the pH at various different values during the fermentation, but the results obtained were poorer than when the pH was uncontrolled.

In attempting to explain the poor production of citric acid from whey permeate when compared to other substrates, Hossain et al.[123] suggested as possible causes the strain of organism used, the nature of the sugar source, and the other nutrients present. Since it is known that the presence of certain trace elements in the fermentation medium can adversely affect citric acid production by *A. niger*,[125] Hossain performed experiments using decationized permeate.[126] The permeate was passed through a cation-exchange resin and subsequently supplemented with various trace elements at selected concentrations. Despite extensive experimentation, no conditions were found which gave an improvement over untreated permeate.

It is now known that lactose is intrinsically a poor substrate for citric acid production. This was demonstrated by Hossain et al.,[127] who used a synthetic medium to compare citric acid production from different sugars. Citric acid concentrations of 53, 31, 23, and 5 g/ℓ were obtained from sucrose, glucose, fructose, and lactose, respectively. No citric acid was produced from galactose despite the fact that this is a readily utilizable carbon source. The poor production from lactose was probably due, at least in part, to the galactose moiety of the disaccharide, although no free galactose was detected in the fermentation broth. Interestingly, citric acid production from whey permeate was superior to that from lactose in the synthetic medium.

It appears, therefore, that the main obstacle to using whey permeate as a substrate for this fermentation process is the inability of the organism to produce citric acid from galactose. In an attempt to solve this problem, Hossain tried to isolate mutant strains which possess this ability, but failed to do so despite intensive efforts.[126] Another possible approach to the problem is to use an organism other than *A. niger*. Here, yeasts of the genera *Candida* and *Saccharomycopsis,* which produce high yields of citric acid from paraffins,[118] may perform well on lactose. Abou-Zeid et al.[128] used a mixture of cheese whey and date seed hydrolysate for citric acid production by *C. lipolytica*. Product concentrations of up to 25 g/ℓ were achieved, but unfortunately no data were reported on the use of whey alone, or on lactose utilization when present in the mixture with data seed hydrolysate. Nevertheless, in view of the problems encountered with *A. niger*, these yeasts appear worthy of further study.

XIII. ACETIC ACID (VINEGAR)

The production of vinegar is a centuries old process, accomplished by the oxidation of ethanol using the bacterium *Acetobacter*. The ethanol is almost invariably obtained as a product of alcoholic fermentation using yeast, and the type of vinegar subsequently obtained is named according to the original raw material, e.g., malt vinegar, wine vinegar, etc. Whey vinegar may be produced by the alcoholic and subsequent acetic fermentation of whey. The basic process of vinegar production has been well documented and need not be described in detail here.[129-131] For whey vinegar, it may be preferable to concentrate the whey prior to fermentation using lactose-fermenting yeasts, so that a suitably high (50 to 60 g/ℓ) ethanol concentration is presented to the *Acetobacter*. Otherwise, the fermentation technologies that may be employed for the alcoholic fermentation are similar to those for ethanol production. After removal of the yeast cells, the acetification stage is conducted by submerged aerobic fermentation in a Frings Acetator,[131] and acetic acid yields of up to 95% may be obtained. Given a suitably high initial ethanol concentration, this can represent an acetic acid concentration of 150 g/ℓ. At present, whey vinegar seems to have a very limited market, although Short has mentioned its production for commercial sale in France and Switzerland, and for use in the production of salad dressings in the U.S.[6]

Glacial acetic acid is an important chemical feedstock that is almost invariably prepared by chemical means. However, Ebner and Follman have described a fermentation process currently operating in Turkey and producing 2 ton of glacial acetic acid per day from molasses.[131] The process involves alcoholic fermentation of diluted molasses followed by separation of purified ethanol by rectification. Diluted ethanol is then treated with *Acetobacter* and the acetic acid is extracted from the broth using ethyl acetate. With the ever-increasing costs of the chemicals used for synthetic acetic acid, a potentially strong market exists for the fermentation product, and whey may prove to be an economical raw material.

XIV. AMINO ACIDS

Several amino acids can be produced by fermentation using appropriate bacteria growing on glucose- or sucrose-containing substrates.[132] Commercial processes are apparently exclusive to Japanese companies. The major use of these products is in supplementation of animal or human foods, although monosodium glutamate also finds application as a flavoring agent. The commercial processes typically use a molasses or glucose syrup substrate, and the bacteria are usually auxotrophic strains which have been isolated following extensive mutation programs. Very little work has been performed outside Japan on these fermentation processes, and whey has, so far, received only scant interest as a substrate. However, the production of lysine, using *Brevibacterium lactofermentum* ATCC 21086, and threonine, using *Escherichia coli* ATCC 21151, have been investigated by Ko and Chipley.[133] The substrates studied were sweet whey permeate, acid-hydolyzed permeate, and enzyme-hydrolyzed permeate. Lysine production from unhydrolyzed permeate was minimal, since the organism failed to use lactose, while using hydrolyzed permeate glucose, but not galactose, was used. The maximum lysine production observed was 3.3 g/ℓ using acid-hydrolyzed permeate supplemented with yeast extract (2 g/ℓ). This represented a yield of only 7% based on glucose utilized. For threonine production, a maximum concentration of 3.6 g/ℓ, representing a yield of 19% based on lactose utilized, was observed when using unhydrolyzed permeate supplemented with yeast extract and ammonium sulfate. The authors stated that the decrease in pH observed during fermentation retarded threonine production, but no experiments were conducted using pH control.

Compared to the commercial production processes for amino acids, the results obtained using a whey permeate substrate are poor. Lysine production from cane molasses can be up to 44 g/ℓ (yield 40%), while threonine production from glucose is reported as 18 g/ℓ (yield 18%).[132] One possible reason for this is that since the organisms used are auxotrophic, whey permeate may be a poor source of the required nutrients. However, a more likely reason may be the difficulty with which the organisms employed utilize lactose and galactose. Thus, the true potential of whey permeate as a substrate for amino acid production may not be recognized until appropriate organisms are developed to cope with these particular sugars.

XV. MICROBIAL POLYSACCHARIDES

Polysaccharide hydrocolloids obtained from plants and seaweed have been used for many years in the food, petroleum and textile industries. Their commercial usefulness is based on their ability to alter the rheological properties of water. Extracellular polysaccharides produced by microorganisms are also now commercially available, and their use has grown as their superior chemical and physical properties have been recognized.[134] Coupled to this is the regularity of their supply via conventional fermentation processes. The first of such microbial polysaccharides to be produced commercially was xanthan gum, using the bacterium *Xanthomonas campestris* growing on a glucose substrate. Its initial use was in the food industry where it acts as a stabilizer and thickening agent, but a more important use is in the recovery of crude oil. The flow characteristics of xanthan, coupled with its pH stability, give it a technical advantage over other polymers in drilling muds. Much effort is being devoted to its use in the recovery of oil from oil strata after conventional recovery methods have been utilized.[134]

The feasibility of using whey permeate as the substrate for xanthan gum production by *X. campestris* has been demonstrated by Charles and Radjai.[135] In their experiments, the lactose was hydrolyzed using a fungal lactase prior to the fermentation, but the necessity for this is not clear. The fermentation is performed under aerobic conditions, at 28°C and pH 7, and equipment should be designed to maximize oxygen transfer, particularly towards the end of the process when the broth becomes quite viscous.[135] Using cheese whey permeate supplemented with small amounts of phsophate and magnesium ions, a xanthan gum concentration of 35 g/ℓ, representing a yield of 85% based on assimilable sugar, has been achieved after 90 hr of fermentation.[135] These results are at least as good as those obtained using conventional media. During the fermentation, glucose and galactose were utilized simultaneously, although glucose at a faster rate.

Another microbial polysaccharide of interest is pullulan, which is produced commercially in Japan from a hydrolyzed starch substrate.[136] The interest stems from the ability of pullulan to form strong, resilient films and fibers, and the ease with which it can be molded into shapes.[134] Le Duy et al.[136] adapted a strain of the pullulan-producing fungus, *Aureobasidium pullulans*, for growth on lactose, with the aim of using whey as a substrate for pullulan production. After adaptation, the organism produced a pullulan concentration of 10.5 g/ℓ from a semisynthetic medium containing lactose (50 g/ℓ). Despite these encouraging results there are, as yet, no reports available using a whey substrate. Studies on the optimum pH for the fermentation have revealed that whereas fungal growth is best at pH 2 to 3, pullulan synthesis is favored at pH 4 to 5.5. This raises the possibility of a two-stage process where nonsterile whey could be used due to the low pH value in the first stage.[136] Le Duy et al.[136] also used the interesting approach of coculturing the fungus *Ceratocystis ulmi* with *A. pullulans*, the purpose of which was to hydrolyze the lactose to glucose and galactose. A 50% increase in pullulan production was observed, but no details were presented with regard to the sugar hydrolysis. However, it does appear that hydrolyzed lactose provides a better substrate, although it is debateable as to whether coculture provides a more favorable approach than pretreatment with isolated lactase enzyme.

Other microbial polysaccharides of commercial interest include dextran and polytran.[134]

Like xanthan gum and pullulan, these are high-value products for which the market appears to be growing rapidly. A very recent report has indicated that whey will be used as the substrate for commercial production of polylevulan gum for food application.[137]

XVI. LACTASE AND CELLULASE

The use of lactase (β-galactosidase) for the hydrolysis of lactose finds application in several areas of the food industry as well as potential application in the treatment of whey.[14,17] Many microorganisms are known to produce lactase, but preparations from yeasts or filamentous fungi appear to find widest commercial application.[138] Suitable sources of yeast lactase include *Saccharomyces fragilis (Kluyveromyces fragilis), S. lactis,* and *Candida pseudotropicalis.*[139] Among the fungi, strains of *Asperillus niger, A. oryzae,* and *Neurospora crassa* represent useful sources,[17,140] while potentially useful bacteria include *Lactobacillus helveticus* and *Streptococcus thermophilus.*[20,140]

Production processes for the yeast lactase generally involve growing the organism on whey under aerobic conditions for 24 to 48 hr. Since the lactase is intracellular, the cells must be harvested and extracted prior to purification of the enzyme. The organism used must be chosen with care as there are important strain differences within a given species.[141] Mahoney et al.[141] reported that when *K. fragilis* was grown on sweet whey at a lactose concentration of 150 g/ℓ, maximum enzyme yield was obtained at the beginning of the stationary phase of growth, and that it was unnecessary to add any supplementary nutrients. In contrast, Wendorff et al.[142] reported that yields were improved when the whey was supplemented with either yeast extract (5 g/ℓ) or corn-steep liquor (5 g/ℓ). de Bales and Castillo,[22] working with *C. pseudotropicalis* in whey, reported that supplementation with yeast extract (1.5 g/ℓ), $(NH_4)_2SO_4$ (1 g/ℓ) and KH_2PO_4 (0.5 g/ℓ) was necessary of maximum enzyme production.

Lactase can be extracted from yeast by various methods, including toluene autolysis, homogenization with glass beads, heat drying, and acetone powder extraction.[141] Toluene autolysis has been described as most satisfactory for *K. fragilis,*[141] while for *C. pseudotropicalis* chloroform gave best results.[22] The purification and physicochemical properties of the *K. fragilis* enzyme have been described by Mahoney and Whitaker.[143]

The production of lactase by *S. thermophilus* grown in deproteinized cheese whey has been described by Ramana Rao and Dutta.[20] The optimum fermentation conditions were pH 7.0 and 40°C. Supplementation of the whey with proteose peptone and corn-steep liquor proved stimulatory to enzyme production, but phosphate or metal ions had no effect.

Production processes for fungal lactase have the advantage that since the enzyme is extracellular it is unnecessary to disrupt the cells prior to its extraction. Further, a lactose-containing medium is not necessary for the production of the fungal enzyme. However recent studies into the production of cellulase by *Trichoderme reesii* have demonstrated the usefulness of lactose as a substrate for this process, and, furthermore, lactase may be obtained as a by-product. The interest in cellulase derives from its use in cellulose hydrolysis prior to using cellulose-containing materials as fermentation substrates for, e.g., ethanol production. The cost of producing the cellulase enzyme makes a large contribution to the overall cost of the cellulose-based fermentation processes, and hence much effort is being devoted to reducing this cost. The cellulase of *T. reesii* is an inducible enzyme, but lactose can serve as the inducer compound.[144] Since a cheap substrate would be beneficial to the production costs, it is possible that whey could be used for this purpose. Furthermore, whey also has the advantage of being less viscous than a cellulose substrate and so would be more suitable for novel fermentation technologies. It has been demonstrated that although lactose is not as good a substrate as cellulose in terms of cellulase enzyme production, the raw material costs of a process based on the former could be less than half those based on the latter.[145,146]

Frein et al.,[147] working in continuous culture with immobilized cells of *T. reesii* on a lactose medium, demonstrated increased productivity of not only cellulase but of all extracellular proteins. Subsequently, Castillo et al.[148] used continuous culture on a lactose medium for lactase production from *T. reesii*. The enzyme is extracellular, and maximum productivity was observed at a dilution rate of 0.027^{-1}, pH 5, and 28°C. Studies on the enzyme demonstrated maximum activity at 63°C and pH 4.6, properties which would make it suitable for hydrolysis of lactose in acid whey.

XVII. GIBBERELLIC ACID

This plant growth hormone, which has wide application in agriculture and the brewing industry, is produced commercially by fungal fermentation using selected strains of *Fusarium moniliforme (Gibberella fujikuroi)*. The fermentation process using conventional media has been well reviewed by Jeffreys.[149] Like many other secondary metabolites, gibberellic acid production is favored in a substrate which is limited in nitrogen, thus suggesting the possible use of whey permeate. The feasibility of this substrate has been demonstrated by Maddox and Richert and Gohlwar et al., who, using different strains of organism, reported gibberellic acid production of 0.5 to 0.6 g/ℓ using unsupplemented permeate as substrate.[71,150,151] Supplementation of the permeate with magensium ions (10 mg/ℓ) caused a 30% increase in production,[150] but addition of nitrogen or phosphate proved detrimental.[71,151] A temperature range of 27 to 30°C and a pH range of 3.5 to 5.5 are considered optimal for the process.[151] Since commercial processes typically use initial sugar concentrations in excess of 100 g/ℓ, then for whey permeate to be a competitive substrate it may require supplementation with lactose or concentration prior to fermentation. As yet, no reports are available describing experiments with such substrates. One possible problem using concentrated permeate may be excess fungal growth leading to problems of oxygen transfer during the process.

XVIII. VITAMINS

Vitamin B$_{12}$ (cyanocobalamin) is produced commercially by fermentation processes, and finds use as a supplement in animal feedstuffs and in the treatment of pernicious anemia in man.[152] The world market is estimated at 10,000 kg/year.[153] Originally, the vitamin was obtained as a by-product during poduction of streptomycin by *Streptomyces griseus* or of chlortetracycline by *S. aureofaciens,* but the concentration in the broth did not usually exceed 2 mg/ℓ.[152] As demand grew, fermentations were designed where vitamin B$_{12}$ is the main product. The organisms used are typically *Propionibacterium freudenreichii, P. shermanii,* or *Pseudomonas denitrificans.*[153] Whatever the strain and culture conditions, a mixture of cobalamins is always obtained. Since the isolation of each distinct substance would be impractical, extraction from the culture medium is generally performed after transformation of all cobalamins into cyanocobalamin by treatment with cyanide.[153] As an alternative to extraction, the entire culture can be fed directly to animals.[154]

Florent and Ninet have described a typical production process using propionibacteria growing on glucose or molasses.[153] The medium must contain sufficient cobalt ions and betaine to ensure satisfactory vitamin production. For some strains of organism 5,6-dimethylbenzimidazole (DBI) must also be provided as a precursor compound. The process is generally run in two stages, the first anaerobic to promote bacterial growth and sugar utilization, the second aerobic to stimulate vitamin B$_{12}$ production. When DBI is added as a precursor, it is usually added during the second stage. Temperature and pH are controlled at 30°C and 6.5 to 7.0, respectively, and vitamin B$_{12}$ concentrations of up to 40 mg/ℓ are produced. When using *Pseudomonas* species the fermentation is conducted under aerobic conditions throughout, and vitamin production of up to 60 mg/ℓ has been reported.[153]

Both whey and whey permeate have been investigated as substrates for the process using various strains of *P. shermanii*.[154-160] Using cheese whey, vitamin B$_{12}$ production has rarely exceeded 20 mg/ℓ, and with permeate only 5 mg/ℓ has been observed. However, this relatively poor production may be a reflection of the strains of organism used rather than the substrate. In all cases, lactose was essentially exhausted from the substrate. The course of the fermentation using cheese whey is similar to that using other substrates, and it is considered essential that the product formation stage be conducted under aerobic rather than anaerobic conditions.[156] In this case, the addition of DBI appears to be unnecessary, at least with the particular strain of organism used. Supplementation of the whey with yeast extract is stimulatory to vitamin production, and addition of cobalt ions at 5 ppm appears adequate.[155-156]

Using whey permeate as substrate, additional supplementary nutrients are required. Thus, in addition to cobalt ions (5 mg/ℓ) and yeast extract (5 g/ℓ), ferrous sulfate (5 mg/ℓ), diammonium hydrogen phosphate (0.5 g/ℓ), betaine (5 g/ℓ) and glutamic acid (0.5 g/ℓ) proved stimulatory to product formation.[160] Otherwise, the fermentation proceeded in a similar manner when using other substrates.

The relative efficiencies of glucose and lactic acid as carbon sources for *P. freudenreichii* have been compared, and it has been shown that although glucose is more efficient as an energy source, lactic acid is a better source for vitamin B$_{12}$ production.[152] Consequently, a two-stage process was described for the production of the vitamin from whey. Initially, the lactose was converted to lactic acid using *Lactobacillus casei*, then *P. freudenreichii* utilized the lactic acid in the second stage of the process.

Riboflavin is another vitamin that can be produced by fermentation processes. However, although it is produced commercially using this route by at least one manufacturer, it is also produced by chemical synthesis.[161] Among the first fermentation processes employed was the use of *Clostridium acetobutylicum* grown on whey.[161] Here, the process was primarily employed for solvents production, the riboflavin being a by-product remaining in the fermentation solids. Today, however, the industrial process employs the fungus *Eromothecium ashbyii* or *Ashbyii gossypii* which can accumulate riboflavin concentrations of up to 7 g/ℓ in the medium.[161] There appear to be no reports describing the use of whey as a substrate for these oganisms.

XIX. BEVERAGES

For many years whey has provided the base for a wide variety of beverages, both alcoholic and nonalcoholic. However, it is difficult to determine the market potential or commercial success of any of these products. Excellent reviews of the various beverages have been provided by Holsinger et al.[162] and Short,[6] and these should be referred to for details of individual products. Broadly, the beverages can be categorized as follows: (1) nonalcoholic products, based on whole or deproteinated whey, and flavored with, e.g., fruit juice; (2) low-alcohol products, where the whey is supplemented with, e.g., sucrose, which is then fermented using *Saccharomyces cerevisiae*. Here, the supplementary sugar is fermented but not the lactose. Such products may be flavored with fruit juices; (3) low-alcohol products where the whey is fermented using *S. lactis*, or, following treatment with a lactase, using *S. cerevisiae*. In both cases, supplementary sugars may be added; (4) whey wine, where concentrated whey is fermented using *S. lactis*,[163] or hydrolyzed, concentrated whey is fermented using *S. cerevisiae*, possibly with the addition of grape juice.[164]

The range of possible beverages based on whey seems enormous, but commercial success may well depend on selecting an appropriate niche in the marketplace.

ACKNOWLEDGMENT

Dr. V. F. Larsen would like to acknowledge that this manuscript was in part prepared while employed at the Biotechnology Dept., Massey University, Palmerston North, New Zealand.

REFERENCES

1. **Zall, R. R., Kuipers, A., Muller, L. L., and Marshall, K. R.,** Trends in whey processing, *N.Z. J. Dairy Sci. Technol.,* 14, 79, 1979.
2. **Anon.,** Effluent technology, in *New Zealand Dairy Research Institute Report,* 1984, 7.
3. **Kavanagh, J. A.,** Production of crude lactose from ultrafiltration permeate, *N.Z. J. Dairy Sci. Technol.,* 10, 132, 1975.
4. **Short, J. L. and Doughty, R. K.,** Demineralization of deproteinated wheys by electrodialysis, *N.Z. J. Dairy Sci. Technol.,* 12, 156, 1977.
5. **Hargrove, R. E., McDonough, F. E., LaCroix, D. E., and Alford, J. A.,** Production and properties of deproteinized whey powders, *J. Dairy Sci.,* 59, 25, 1976.
6. **Short, J. L.,** Prospects for the utilization of deproteinated whey in New Zealand — a review, *N.Z. J. Dairy Sci. Technol.,* 13, 181, 1978.
7. **Archer, R. H., Larsen, V. F., and McFarlane, P. N.,** Comparison of three high rate anaerobic digester design for the treatment of whey: preliminary results, presented at Institute of Professional Engineers of New Zealand Annual Meeting, Hamilton, February, 1982.
8. **Nickerson, T. A.,** Lactose, in *Byproducts from Milk,* 2nd ed., Webb, B. H. and Whittier, E. O., Eds., AVI, Westport, Conn., 1970, 356.
9. **Hobman, P. G.,** Lactose — processes and products, *J. Dairy Sci.,* 67, 2630, 1984.
10. **Ruiz, L. P., Gurnsey, J. C., and Short, J. L.,** Reduction of lactic acid, nonprotein nitrogen, and ash in lactic acid whey by *Candida ingens* culture, *Appl. Environ. Microbiol.,* 35, 771, 1978.
11. **Cerbulis, J.,** Application of Steffen process and its modifications to recovery of lactose and protein in whey, *J. Agric. Food Chem.,* 21, 255, 1973.
12. **Kwon, S. Y., Bernhard, R. A., and Nickerson, T. A.,** Recovery of lactose from aqueous solutions: precipitation with magnesium chloride and sodium hydroxide, *J. Dairy Sci.,* 64, 396, 1981.
13. **Quickert, S. C. and Bernhard, R. A.,** Recovery of lactose from aqueous solution using group IIA metal chlorides and sodium hydroxide, *J. Food Sci.,* 47, 1705, 1982.
14. **Gregory, K.,** New dairy processes using lactase, *Food Proc. Ind.,* 49(12), 44, 1980.
15. **Coton, S. G.,** The utilization of permeates from the ultrafiltration of whey and skim milk, *J. Soc. Dairy Technol.,* 33, 89, 1980.
16. **Ennis, B. M.,** Lactose hydrolysis in New Zealand, *N.Z. J. Dairy Sci. Technol.,* 17, A21, 1982.
17. **Prenosil, J. E., Stucker, E., Hediger, T., and Bourne, J. R.,** Enzymatic whey hydrolysis in the pilot plant "Lactohyd", *Bio/Technol.,* 441, 1984.
18. **Finocchiaro, T., Olson, N. F., and Richardson, T.,** Use of immobilized lactase in milk systems, *Adv. Biochem. Eng.,* 15, 71, 1980.
19. **MacBean, R. D.,** Lactose crystallization and lactose hydrolysis, *N.Z. J. Dairy Sci. Technol.,* 14, 113, 1979.
20. **Ramana Rao, M. V. and Dutta, S. M.,** Production of beta-galactosidase from *Streptococcus thermophilus* grow in whey, *Appl. Environ. Microbiol.,* 34, 185, 1977.
21. **Griffiths, M. W. and Muir, D. D.,** Properties of a thermostable β-galactosidase from a thermophilic *Bacillus:* comparison of the enzyme activity of whole cells, purified enzyme and immobilized whole cells, *J. Sci. Food Agric.,* 29, 753, 1978.
22. **deBales, S. A. and Castillo, F. J.,** Production of lactase by *Candida pseudotropicalis* grown in whey, *Appl. Environ. Microbiol.,* 37, 1201, 1979.
23. **Greenberg, N. A. and Mahoney, R. R.,** Immobilisation of lactase (β-galactosidase) for use in dairy processing: a review, *Proc. Biochem.,* 16(1), 2, 1981.
24. **Brodelius, P.,** Industrial applications of immobilized biocatalysts, *Adv. Biochem. Eng.,* 10, 75, 1978.
25. **Pitcher, W. H.,** Design and operation of immobilized enzyme reactors, *Adv. Biochem. Eng.,* 10, 1, 1978.
26. **Dicker, R.,** Whey hydrolysis, a "sweet" breakthrough, *Dairy Ind. Int.,* 47(4), 19, 1982.
27. **MacBean, R. D., Hall, R. J., and Willman, N. J.,** Heterogenous acid catalysed hydrolysis of lactose with cation exchange resins, *Aust. J. Dairy Technol.,* 34, 53, 1979.

28. **Poutanen, K., Linko, Y., and Linko, P.,** Treatment of hydrolysed whey lactose syrup with immobilized glucose isomerase for increased sweetness, *Milchwissenschaft, 33,* 435, 1978.

29. **Buswell, A. M., Boruff, C. S., and Wiesman, C. K.,** Anaerobic stabilisation of milk waste, in *Anaerobic Fermentations,* Buswell, A. M. and Hatfield, W. D., Eds., State of Illinois, Urbana, 1939, 146.

30. **Parker, J. G. and Skerry, G. P.,** Treatment of milk processing wastes by methane fermentation, lagoon and activated sludge processes, presented at Third Natl. Chem. Eng. Conf., Mildura, Victoria, Australia, August 20—23, 1975.

31. **Parker, J. G. and Lyons, B. J.,** Factors influencing the treatment of food processing wastes by anaerobic-aerobic lagoon systems, *Prog. Water Technol.,* 11, 377, 1979.

32. **Follmann, H. and Markl, H.,** PH-statisches Verfahren zur kontinuierlichen anaeroben Vergarung von hochbelasteten Abwassern zu Methan. II. Experimentelle Untersuchungen und Verfizierung des Konzeptes am Beispiel der Molke, *Milchwissenschaft,* 34, 393, 1979.

33. **Callander, I. J. and Barford, J. P.,** Recent advances in anaerobic digestion technology, *Proc. Biochem.,* 18(4), 24, 1983.

34. **Watson, D. A.,** Pilot scale anaerobic digestion trials on dairy effluents using the contact porcess, presented at Int. Dairy Fed. Seminar Dairy Effluents, Killarney, Ireland, April 19—22, 1983.

35. **Frostell, B.,** Anamet anaerobic-aerobic treatment of concentrated wastewaters, *Proc. 36 Int. Waste Treatment Conf.,* Purdue, 1981, 269.

36. **Callander, I. J. and Barford, J. P.,** Anaerobic digestion of power ethanol stillage and whey effluent using flocculated microorganisms in a continuous up-flow fermenter, *Proc. 14 New Zealand Biotechnol Conf.,* Palmerston North, New Zealand, 1982, 166.

37. **Archer, R. H., Larsen, V. F., and McFarlane, P. N.,** Anaerobic digestion of whey permeate: a comparison of three reactor designs, presented at Third Int. Symp. Anaerobic Digestion, Boston, Mass., August 14—19, 1983.

38. **Archer, R. H.,** personal communication, 1984.

39. **van den Berg, L. and Kennedy, K. J.,** Dairy waste treatment with anaerobic stationary fixed film reactors, *Proc. IAWPR Specialised Seminar,* Copenhagen, Denmark, June 1982, 229.

40. **Barnett, J. W.,** Anaerobic Filtration of Waste Waters Arising from the Production of Baker's Yeast, Ph.D. thesis, Massey University, Palmerston North, New Zealand, 1985.

41. **Young, J. C. and McCarty, P. L.,** The anaerobic filter for waste treatment, *Proc. 22nd Ind. Waste Conf.,* Purdue University, May 1967, 559.

42. **Hakansson, H., Frostell, B., and Norrman, J.,** Anaerobic treatment of whey and whey permeate in biological filter, *Report B-406,* Institute for Vatten och luftvardsforskning, Stockholm, Sweden, 1977.

43. **Maranan, E. A.** Anaerobic Filtration of Whey, Dip. Technol. dissertation, Massey University, Palmerston North, New Zealand, 1981.

44. **de Haast, J., Britz, T. J., Novello, J. C., and Lategan, P. M.,** Anaerobic digestion of cheese whey using stationary fixed-bed reactor, *N.Z. J. Dairy Sci. Technol.,* 18, 261, 1983.

45. **Boening, P. H. and Larsen, V. F.,** Anaerobic fluidized bed whey treatment, *Biotechnol. Bioeng.,* 24, 2539, 1982.

46. **Switzenbaum, M. S. and Danskin, S. C.,** Anaerobic expanded bed treatment of whey, *Proc. 36th Ind. Waste Conf.,* Purdue, May 1981, 414.

47. **Sutton, P. M. and Li, A.,** Anitron and oxitron systems: High rate anaerobic and aerobic biological treatment systems for industry, *Proc 36th Waste Conf.,* Purdue, May 1981, 665.

48. **Boening, P. and Larsen, V. F.,** Fluidised bed anaerobic digester, *Proc. 9th Australasian Conf. Chem. Eng.,* Christchurch, New Zealand, August 30 to September 4, 1981, 527.

49. **Titchner, A. L.,** The Brazilian alcohol program and some implications for New Zealand, presented at Inst. Professional Eng. Annu. Meet., Dunedin, New Zealand, February 1980.

50. **Yoo, B. W.,** Effect of Certain Factors on Cell Mass and Ethanol Production by Lactose Fermenting Yeasts, Ph.D. thesis, University of Maryland, College Park, 1974.

51. **Browne, H. H.,** Ethyl alcohol from fermentation of lactose in whey, *Ind. Eng. Chem. News Ed.,* 19, 1272, 1941.

52. **Rogosa, M., Browne, H. H., and Whittier, E. O.,** Ethyl alcohol from whey, *J. Dairy Sci.,* 30, 263, 1947.

53. **Moulin, G., Guilliaume, M., and Galzy, P.,** Alcohol production by yeast in whey ultrafiltrate, *Biotechnol. Bioeng.,* 22, 1277, 1980.

54. **Janssens, J. H., Bernard, A., and Bailey, R. B.,** Ethanol from whey: continuous fermentation with cell recycle, *Biotechnol. Bioeng.,* 26, 1, 1984.

55. **Boontanjai, C.,** Tower Fermentation of Whey Permeate and Sucrose Enriched Whey Permeate to Ethanol, Ph.D. thesis, Massey University, Palmerston North, New Zealand, 1983.

56. **Linko, Y., Jalanka, H., and Linko, P.,** Ethanol production from whey with immobilized living yeast, *Biotechnol. Lett.,* 3, 263, 1981.

57. **O'Leary, V. S., Green, R., Sullivan, B. C., and Holsinger, V. H.,** Alcohol production by selected yeast strains in lactase-hydrolyzed acid whey, *Biotechnol. Bioeng.,* 19, 1019, 1977.
58. **Bernstein, S., Tzeng, C. H., and Sisson, D.,** The commercial fermentation of cheese whey for the production of protein and/or alcohol, *Biotechnol. Bioeng. Symp.,* 7, 1, 1977.
59. **Gavel, J. and Kosikowski, F. V.,** Improving alcohol fermentation in concentrated ultrafiltration permeates of cottage cheese whey, *J. Food Sci.,* 43, 1717, 1978.
60. **Burgess, K. J. and Kelly, J.,** Alcohol production by yeast in concentrated ultrafiltration permeate from cheddar cheese whey, *Ir. J. Food Sci. Technol.,* 3, 1, 1979.
61. **Izaguirre, M. E. and Castillo, F. J.,** Selection of lactose fermenting yeasts for ethanol production from whey, *Biotechnol. Lett.,* 4, 257, 1982.
62. **Larsen, V. F.,** Alcohol recovery, in *Yeast Biotechnology,* Berry, D. R., Stewart, G. G., and Russel, I., Eds., Allen & Unvin, London, in press.
63. **Howell, M. J.,** High-grade ethanol from whey, *Chem. Ind. Lab. Manage.,* (Auckland), 15(5), 4, 1981.
64. **Reesen, L.,** Alcohol production from whey, *Dairy Ind. Int.,* 43(1), 9, 1978.
65. **Dillon, J. A.,** The Immobilization of *Kluyveromyces fragilis* and *Saccharomyces cerevisiae* in Polyacrylamide Gel, M. Technol. (Biotechnol.) thesis, Massey University, Palmerston North, New Zealand, 1980.
66. **O'Leary, V. S., Sutton, S., Bencivengo, M., Sullivan, B., and Holsinger, V. H.,** Influence of lactose hydrolysis and solids concentration on alcohol production by yeast in acid whey ultrafiltrate, *Biotechnol. Bioeng.,* 19, 1698, 1977.
67. **Moulin, G., Boze, H., and Galzy, P.,** Inhibition of alcohol fermentation by substrate and ethanol, *Biotechnol. Bioeng.,* 22, 2375, 1980.
68. **Janssens, J. H., Burris, N., Woodward, A., and Bailey, R. B.,** Lipid-enhanced ethanol production by *Kluyveromyces fragilis, Appl. Environ. Microbiol.,* 45, 598, 1983.
69. **Meyrath, J. and Bayer, K.,** Biomass from whey, in *Economic Microbiology,* Vol. 4, *Microbial Biomass,* Rose, A. H., Ed., Academic Press, London, 1979, 207.
70. **Trevis, T.,** Whey-yeast protein surpasses soymeal as protein source, *Feedstuffs,* June 25, 12, 1979.
71. **Maddox, I. S. and Richert, S. H.,** Use of response surface methodology for the rapid optimization of microbiological media, *J. Appl. Bacteriol.,* 43, 197, 1977.
72. **Woodbine, M.,** Microbial fat: micirorganisms as potential fat producers, *Prog. Ind. Microbiol.,* 1, 181, 1959.
73. **Ratledge, C.,** Microbial oils and fats: an assessment of their commercial potential, *Prog. Ind. Microbiol.,* 16, 119, 1982.
74. **Hammond, E. G., Glatz, B. A., Choi, Y., and Teasdale, M. T.,** Oil production by *Candida curvata* and extraction composition, and properties of the oil, *AOCS Monogr.,* 9, 171, 1981.
75. **Wix, P. and Woodbine, M.,** Mycological synthesis of fat from whey. I. Preliminary studies with stationary cultures, *J. Appl. Bacteriol.,* 22, 14, 1959.
76. **Wix, P. and Woodbine, M.,** Mycological synthesis of fat from whey. II. Comparative studies with shaken and stationary cultures using selected moulds, *J. Appl. Bacteriol.,* 22, 175, 1959.
77. **Mickle, J. B., Smith, W., Halter, D., and Knight, S.,** Performance and morphology of *Kluyveromyces fragilis* and *Rhodotorula gracilis* grown in cottage cheese whey, *J. Milk Food Technol.,* 37, 481, 1974.
78. **Moon, N. J. and Hammond, E. G.,** Oil production by fermentation of lactose and the effect of temperature on the fatty acid composition, *J. Am. Oil Chem. Soc.,* 55, 683, 1978.
79. **Moon, N. J., Hammond, E. G., and Glatz, B. A.,** Conversion of cheese whey and whey permeate to oil and single-cell protein, *J. Dairy Sci.,* 61, 1537, 1978.
80. **Davies, J. and Gordon, T.,** Continuous production of yeast oil from whey, in *Proc. 6th Australian Biotechnol. Conf.,* University of Queensland, Brisbane, Australia, August, 1984, 127.
81. **Floetenmeyer, M. D., Glatz, B. A., and Hammond, E. G.,** Fermentation of cheese whey permeate with the oleaginous yeast *Candida curvata, J. Dairy Sci.,* 67(Suppl. 1), 56, 1984.
82. **Gapes, J. R.,** The Acetone-Butanol-Ethanol Fermentation: Preliminary Studies on Some Practical Aspects, M. Technol. thesis, Massey University, Palmerston North, New Zealand, 1982.
83. **Prescott, S. C. and Dunn, C. G.,** *Industrial Mircobiology,* 3rd ed., McGraw-Hill, New York, 1959, 250.
84. **Spivey, M. J.,** The acetone/butanol/ethanol fermentation, *Process Biochem.,* 13(11), 2, 1978.
85. **Maddox, I. S.,** Production of ethanol and n-butanol from hexose/pentose mixtures using consecutive fermentations with *Saccharomyces cerevisiae* and *Clostridium acetobutylicum, Biotechnol. Lett.,* 4, 23, 1982.
86. **Maddox, I. S.,** Production of n-butanol from whey filtrate using *Clostridium acetobutylicum* NCIB 2951, *Biotechnol. Lett.,* 2, 493, 1980.
87. **Welsh, F. W. and Veliky, I. A.,** Production of acetone-butanol from acid whey, *Biotechnol. Lett.,* 6, 61, 1984.
88. **Maddox, I. S., Gapes, J. R., and Larsen, V. F.,** Production of n-butanol from whey ultrafiltrate, in *Proc. 9th Australasian Conf. Chemical Eng.,* Chemical Engineering Group, Wellington, New Zealand, 1981, 535.

89. **Schoutens, G. H., Nieuwenhuizen, M. C. H., and Kossen, N. W. F.,** Butanol from whey ultrafiltrate: batch experiments with *Clostridium beyerinckii* LMD 27.6, *Appl. Microbiol. Biotechnol.,* 19, 203, 1984.
90. **Ewen, M. and Maddox, I. S.,** unpublished data, 1983.
91. **Gapes, J. R., Larsen, V. F., and Maddox, I. S.,** A note on procedures for inoculum development for the production of solvents by a strain of *Clostridium butylicum, J. Appl. Bacteriol.,* 55, 363, 1983.
92. **Krouwel, P. G., Groot, W. J., Kossen, N. W. F., and van der Laan, W. F. M.,** Continuous isopropanol-butanol-ethanol fermentation by immobilized *Clostridium beyerinckii* cells in a packed bed fermenter, *Enzyme Microbiol. Technol.,* 5, 46, 1983.
93. **Monot, F. and Engasser, J. M.,** Continuous production of acetone butanol on an optimized synthetic medium, *Eur. J. Appl. Microbiol. Biotechnol.,* 18, 246, 1983.
94. **Schoutens, G. H., Nieuwenhuizen, M. C. H., and Kossen, N. W. F.,** personal communication, 1984.
95. **Lenz, T. G. and Moreira, A. R.,** Economic evaluation of the acetone-butanol fermentation, *Ind. Eng. Chem. Prod. Res. Devel.,* 19, 478, 1980.
96. **Volesky, B., Mulchandani, A., and Williams, J.,** Biochemical production of industrial solvents (acetone-butanol-ethanol) from renewable resources in *Biochemical Engineering II: Proceedings,* Constantinides, A., Vieth, W. R., and Venkatasubramanian, K., Eds., *N.Y. Academy of Science,* 1981, 205.
97. **Larsen, V. F.,** unpublished data.
98. **Jansen, N. B. and Tsao, G. T.,** Bioconversion of pentoses to 2,3-butanediol by *Klebsiella pneumoniae, Adv. Biochem. Eng. Biotechnol.,* 27, 85, 1983.
99. **Prescott, S. C. and Dunn, C. G.,** *Industrial Microbiology,* 3rd ed., McGraw-Hill, New York, 1959, 399.
100. **Speckman, R. A. and Collins, E. B.,** Microbial production of 2,3-butylene glycol from cheese whey, *Appl. Environ. Microbiol.,* 43, 1216, 1982.
101. **Shazer, W. H. and Speckman, R. A.,** Continuous fermentation of whey permeate by *Bacillus polymyxa, J. Dairy Sci.,* 67(Suppl. 1), 50, 1984.
102. **Lee, H. K.,** Production of 2,3-Butanediol by Fermentation of Whey Permeate, Dip. Technol. dissertation, Massey University, Palmerston North, New Zealand, 1984.
103. **Lee, H. K. and Maddox, I. S.,** unpublished data, 1984.
104. **Miall, L. M.,** Organic acids, in *Economic Microbiology,* Vol. 2, *Primary Products of Metabolism,* Rose, A. H., Ed., Academic Press, London, 1978, 48.
105. **Vick Roy, T. B., Mandel, D. K., Dea, D. K., Blanch, H. W., and Wilke, C. R.,** The application of cell recycle to continuous fermentative lactic acid production, *Biotechnol. Lett.,* 5, 665, 1983.
106. **Anon.,** *Biotechnol. News,* 4(22), 5, 1984.
107. **Whittier, E. O.,** Fermentation products from whey, in *Byproducts from Milk,* Whittier, E. O. and Webb, B. H., Eds., Reinhold, New York, 1950, chap. 3.
108. **Prescott, S. C. and Dunn, C. G.,** *Industrial Microbiology,* 3rd ed., McGraw-Hill, New York, 1959, 304.
109. **Marshall, K. R. and Earle, R. L.,** The effect of agitation on the production of lactic acid by *Lactobacillus bulgaricus* in a casein whey medium, *N.Z. J. Dairy Sci. Technol.,* 10, 123, 1975.
110. **Marshall, K. R.,** The Production of Lactic Acid from Whey by Continuous Culture as a Possible Means of Waste Disposal, Ph.D. thesis, Massey University, Palmerston North, New Zealand, 1972.
111. **Keller, A. K. and Gerhardt, P.,** Continuous lactic acid fermentation of whey to produce a ruminant feed supplement high in crude protein, *Biotechnol. Bioeng.,* 17, 977, 1975.
112. **Setti, D.,** Development of a new technology for lactic acid production from cheese whey, *Proc. 4th Int. Congr. Food Sci. Technol.,* 4, 289, 1974.
113. **Friedman, M. R. and Gadan, E. L.,** Growth and acid production by *Lactobacillus* (L.) *delbruecki* in a dialysis culture system, *Biotechnol. Bioeng.,* 12, 961, 1970.
114. **Stieber, R. W. and Gerhardt, P.,** Dialysis continuous process for ammonium lactate fermentation: simulated and experimental dialysis-feed immobilized cell systems, *Biotechnol. Bioeng.,* 23, 535, 1981.
115. **Linko, P.,** Immobilized live cells, in *Advances in Biotechnology,* Moo-Young, M., Ed., Pergamon Press, Toronto, 1981, 711.
116. **Stenroos, S. L., Linko, Y. Y., and Linko, P.,** Production of *l*-lactic acid with immobilized *Lactobacillus delbrueckii, Biotechnol. Lett.,* 4, 159, 1982.
117. **Vick Roy, T. B., Blanch, H. W., and Wilke, C. R.,** Lactic acid production by *Lactobacillus delbrueckii* in a hollow fiber fermenter, *Biotechnol. Lett.,* 4, 483, 1982.
118. **Kapoor, K. K., Choudhary, K., and Tauro, P.,** Citric acid, in *Prescott and Dunn's Industrial Microbiology,* 4th ed., Reed, G., Ed., AVI, Westport, Conn., 1982, 709.
119. **Smith, J. E., Nowakowska-Waszczuk, A., and Anderson, J. G.,** Organic aicd production by mycelial fungi, in *Industrial Aspects of Biochemistry,* Vol. 1, Spencer, B., Ed., Elsevier, Amsterdam, 1974, 297.
120. **Lockwood, L. B.,** Organic acid production, in *The Filamentous Fungi,* Vol. 1, Smith, J. E. and Berry, D. R., Eds., Edward Arnold, London, 1975, 140.
121. **Sodeck, G., Modl, J., Kominek, J., and Salzbrunn, W.,** Production of citric acid according to the submerged fermentation process, *Process. Biochem.,* 16(5), 9, 1982.

122. **Somkuti, G. A. and Bencivengo, M. M.,** Citric acid fermentation in whey permeate, *Devel. Ind. Microbiol.,* 22, 557, 1981.

123. **Hossain, M., Brooks, J. D., and Maddox, I. S.,** Production of citric acid from whey permeate fermentation using *Aspergillus niger, N.Z. J. Dairy Sci. Technol.,* 18, 161, 1983.

124. *Catalogue of the Culture Collection of the Commonwealth Mycological Institute,* 7th ed., Kew, Surrey, 1975.

125. **Berry, D. R., Chmiel, A., and Al Obaidi, Z.,** Citric acid production by *Aspergillus niger,* in *Genetics and Physiology of Aspergillus,* Smith, J. E. and Pateman, J. A., Eds., Academic Press, London, 1977, 405.

126. **Hossain, M.,** Submerged Citric Acid Fermentation of Whey Permeate by *Aspergillus niger,* Ph.D. thesis, Massey University, Palmerston North, New Zealand, 1983.

127. **Hossain, M., Brooks, J. D., and Maddox, I. S.,** The effect of the sugar source on citric acid production by *Aspergillus niger, Appl. Microbiol. Biotechnol.,* 19, 393, 1984.

128. **Abou-Zeid, A. Z. A., Baghlaf, A. O., Khan, J. A., and Makhashin, S. S.,** Utilization of date seeds and cheese whey in production of citric acid by *Candida lipolytica, Agric. Wastes,* 8, 131, 1983.

129. **Greenshields, R. N.,** Acetic aicd: vinegar, in *Economic Microbiology,* Vol. 2, *Primary Products of Metabolism,* Rose, A. H., Ed., Academic Press, London, 1978, 121.

130. **Ebner, H.,** Vinegar, in *Prescott & Dunn's Industrial Microbiology,* 4th ed., Reed, G., Ed., AVI, Westport, Conn., 1982, chap. 18.

131. **Ebner, H. and Follman, H.,** Acetic acid, in *Biotechnology,* Vol. 3, Rehm, H. J. and Reed, G., Eds., Verlag-Chemie, Weinheim, 1983, chap. 3b.

132. **Kinoshita, S. and Nakayama, K.,** Amino acids, in *Economic Microbiology,* Vol. 2, Rose, A. H., Ed., Academic Press, London, 1978, 209.

133. **Ko, Y. T. and Chipley, J. R.,** Microbial production of lysine and threonine from whey permeate, *Appl Environ. Microbiol.,* 45, 610, 1983.

134. **Wells, J.,** Extracellular microbial polysaccharides — a critical overview, in *Extracellular Microbial Polysaccharides,* Sandford, P. A. and Laskin, A., Eds., A.C.S. Series 45, American Chemical Society, Washington, D.C., 1977, 299.

135. **Charles, M. and Radjai, M. K.,** Xanthan gum from acid whey, in *Extracellular Microbial Polysaccharides,* Sandford, P. A. and Laskin, A., Eds., A.C.S. Series 45, American Chemical Society, Washington, D.C., 1977, 27.

136. **LeDuy, A., Yarmoff, J. J., and Chagraoui, A.,** Enchanced production of pullulan from lactose by adaptation and by mixed culture techniques, *Biotechnol. Lett.,* 5, 49, 1983.

137. **Anon.,** *Biotechnol. News,* 4(22), 6, 1984.

138. **Itoh, T., Suzuki, M., and Adachi, S.,** Production and characterization of β-galactosidase from lactose-fermenting yeasts, *Agric. Biol. Chem.,* 46, 899, 1982.

139. **Harrison, J. A.,** Miscellaneous products from yeast, in *The Yeasts,* Vol. 3, Rose, A. H. and Harrison, J. S., Eds., Academic Press, London, 1970, 529.

140. **Wierzbicki, L. E. and Kosikowski, F. V.,** Lactase potential of various microorganisms grown in whey, *J. Dairy Sci.,* 56, 26, 1973.

141. **Mahoney, R. R., Nickerson, T. A., and Whitaker, J. R.,** Selection of strain, growth conditions, and extraction procedures for optimum production of lactase from *Kluyveromyces fragilis, J. Dairy Sci.,* 58, 1620, 1974.

142. **Wendorff, W. L., Amundson, C. H., and Olson, N. F.,** Nutrient requirements and growth conditions for production of lactase enzyme by *Saccharomyces fragilis, J. Milk Food Technol.,* 33, 451, 1970.

143. **Mahoney, R. A. and Whitaker, J. R.,** Purification and physicochemical properties of β-galactosidase from *Kluyveromyces fragilis, J. Food Sci.,* 43, 584, 1978.

144. **Sternberg, D. and Mandels, G. R.,** Induction of cellulolytic enzymes in *Trichoderme reesii* by sophorose, *J. Bacteriol.,* 139, 761, 1979.

145. **Warzywoda, M., Ferre, V., and Pourquie, J.,** Development of a culture medium for large scale production of cellulolytic enzymes by *Trichoderme reesii, Biotechnol. Bioeng.,* 25, 3005, 1983.

146. **Warzywoda, M., Vandecasteele, J. P., and Pourquie, J.,** A comparison of genetically improved strains of the cellulolytic fungus *Trichoderme reesssi, Biotechnol. Lett.,* 5, 243, 1983.

147. **Frein, E. M., Montenecourt, B. S., and Eveleigh, D. E.,** Cellulase production by *Tricoderme reesii* immobilized on κ-carragheenan, *Biotechnol. Lett.,* 4, 287, 1982.

148. **Castillo, F. J., Blanch, H. W., and Wilke, C. R.,** Lactase production in continuous culture by *Trichoderme reesii* RUT-C30, *Biotechnol. Lett.,* 6, 593, 1984.

149. **Jeffreys, E. G.,** The gibberellin fermentation, *Adv. Appl. Microbiol.,* 13, 283, 1970.

150. **Maddox, I. S. and Richert, S. H.,** Production of gibberellic acid using a dairy waste as the basal medium, *Appl. Environ. Microbiol.,* 33, 201, 1977.

151. **Gohlwar, C. S., Sethl, R. P., Marwaha, S. S., Seghal, V. K., and Kennedy, J. F.,** Gibberellic acid biosynthesis from whey and simulation of cultural parameters, *Enzyme Microbiol. Technol.,* 6, 312, 1984.

152. **Mervyn, L. and Smith, E. L.,** The biochemistry of vitamin B_{12} fermentation, *Prog. Ind. Microbiol.,* 5, 151, 1964.
153. **Florent, J. and Ninet, L.,** Vitamin B_{12}, in *Microbial Technology,* 2nd ed., Peppler, H. J. and Perlman, D., Eds., Academic Press, New York, 1979, 497.
154. **Marawaha, S. S. and Sethi, R. P.,** Utilization of dairy waste for vitamin B_{12} fermentation, *Agric. Wastes,* 9, 111, 1984.
155. **Bullerman, L. B. and Berry, E. C.,** Use of cheese whey for vitamin B_{12} production. I. Whey solids and yeast extract levels, *Appl. Microbiol.,* 14, 353, 1966.
156. **Berry, E. C. and Bullerman, L. B.,** Use of cheese whey for vitamin B_{12} production. II. Cobalt, precursor and aeration levels, *Appl. Microbiol.,* 14, 356, 1966.
157. **Bullerman, L. B. and Berry, E. C.,** Use of cheese whey for vitamin B_{12} production. III. Growth studies and dry-weight activity, *Appl. Microbiol.,* 14, 358, 1966.
158. **Marwaha, S. S., Sethi, R. P., and Kennedy, J. F.,** Influence of 5,6-dimethylbenzimidazole (DMB) on vitamin B_{12} biosynthesis by strains of *Propionibacterium, Enzyme Microbiol. Technol.,* 5, 361, 1983.
159. **Marwaha, S. S., Sethi, R. P., Kennedy, J. F., and Kumar, R.,** Simulation of fermentation conditions for vitamin B_{12} biosynthesis from whey, *Enzyme Microbiol. Technol.,* 5, 449, 1983.
160. **Marwaha, S. S., Sethi, R. P., and Kennedy, J. F.,** Role of amino acids, betaine and choline in vitamin B_{12} biosynthesis by strains of *Propionibacterium, Enzyme Microbiol. Technol.,* 5, 454, 1983.
161. **Perlman, D.,** Microbial process for riboflavin production, in *Microbial Technology,* 2nd ed., Peppler, H. J. and Perlman, D., Eds., Academic Press, New York, 1979, 521.
162. **Holsinger, V. H., Posati, L. P., and de Vilbiss, E. D.,** Whey beverages: a review, *J. Dairy Sci.,* 57, 849, 1974.
163. **Kosikowsky, F. V. and Wzorek, W.,** Whey wine from concentrates of reconstituted acid whey powder, *J. Dairy Sci.,* 60, 1982, 1978.
164. **Roland, J. F. and Alm, W. L.,** Wine fermentations using membrane processed hydrolysed whey, *Biotechnol. Bioeng.,* 17, 1443, 1975.

Chapter 3

ZETA POTENTIAL MEASUREMENTS FOR BIOTECHNOLOGY APPLICATIONS

Ph. Thonart and M. Paquot

TABLE OF CONTENTS

I. INTRODUCTION

Many research workers, industrialists, and politicians would agree that a technological battle has commenced. This situation can be ascribed to the acquisition by industries of biotechnological knowledge. Biotechnology consists of the industrial exploitation of microorganism potential, of animal and plant cells, as well as the subcellular fractions derived from these.

However, biotechnology is not a recent acquisition. The development of agriculture is the result of man learning how best to use the growth and development of plant and animal cells since Neolithic times. If man has managed to make bread, wine, and beer, it is thanks to the fermentative properties of yeast.

The textile industry has long used biological processes for separating the fibers of flax and hemp. It was often very difficult to optimize these techniques, for the mechanisms of life, and even certain forms of life, were still unknown. Today, with the development of knowledge, each phase of the biotechnological process can be optimized (see Figure 1).

This optimization implies of course the use of both biochemical and genetic engineering. In the biochemical processes, we describe the importance of the charge and the cell wall.

Cell activity and the interplay with the environment constitute the primary elements of a microbiological system. The perpetual interaction and interdependence of the total environment and the living cells lead to the maintenance of a balanced existence within each system.

Environmental factors directly affect the metabolism of the living organism, and the actions of the individual cell are genetically controlled.[1] The cell wall region plays an important role in the interplay of cell and environmental factors, especially in mass transfer, flocculation properties, etc.

Figure 2 shows that another possibility of improving biotechnological processes utilizes biotechnological reactors using immobilized microbial cells. The surface properties of microbial cells and carriers are of considerable importance in cell immobilization by adsorption. In the case of immobilized enzymes, the interaction between the insoluble support and the soluble enzyme can be optimized by the charge measurement of the carrier.

II. COLLOIDAL STATE

For several applications, biotechnology, like many other industries, is faced with several problems which are based at least partially on colloidal phenomena. Colloidal systems have been the subject of rather intensive investigation. Historically, two major classes of colloidal systems have been recognized: lyophilic and lyophobic colloids.

Gelatin, starch, and other hydrocolloids show a marked affinity for water of other appropriate solvents. They acquire stability by solvation of the interface, i.e., any degrees of interaction from more physical wetting to the formation of adherent thick layers of oriented solvent molecules.

These layers may display highly viscous or gelated properties. These products are classed as lyophilic colloids. Liquid dispersions of small solid or liquid particles produced by mechanical or chemical means are termed "lyophobic colloids". Lyophobic colloids are stabilized by an electrostatic repulsion between particles arising from ions that are either sorbed onto or dissolved out of the surface. Lyophilic colloids are essentially true solutions and perhaps can be described as macromolecular colloids.[2] The primary property of lyophobic colloids is their great sensitivity to electrolytes. Addition of small amounts of soluble salts to such dispersions generally cause the particles to clump together, or flocculate. Lyophilic colloids, being true solutions, are insensitive to electrolytes and do not flocculate, although they may be precipitated by high salt concentrations.

Thus, one colloid system (lyophilic) is governed by solvation; the other (lyophobic) is

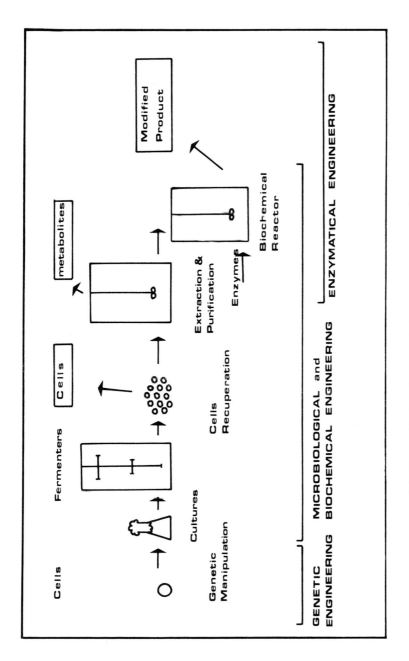

FIGURE 1.   Classical biotechnological processes and enzyme utilization.

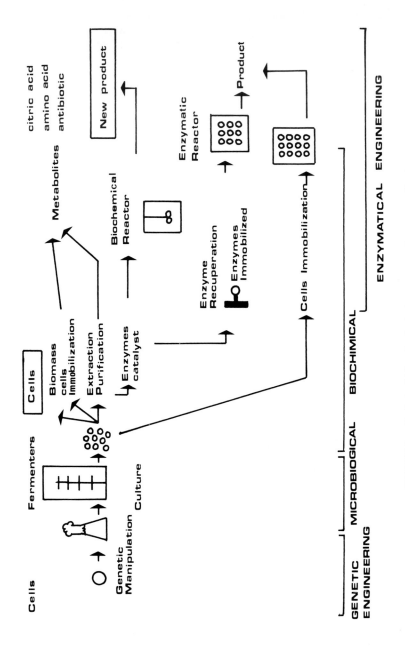

FIGURE 2.   Optimized biotechnological processes and enzyme utilization.

governed by electric charge. In reality, the two mechanisms are not mutually exclusive.[3] They often occur together, even if one predominates. For example, the adsorbed ions that are responsible for electric charge concomitantly provide a solvated layer on the particle surface.

When the particle surfaces obtain an electric charge, they develop effective double layers that depend on the presence of a low concentration of specific ions dissolved in the medium.

The significant properties of the electric double layer which can be experimentally measured is the zeta potential.

III. ZETA POTENTIAL

A number of particles in suspension carry a charge which is negative in the majority of cases.[4] This charge is cancelled by positive ions captured in the surrounding medium and which are evenly distributed following a decreasing pattern at the solid-liquid interface. The positive charges directly adjacent to, and strongly attracted by, the negative surface behave as if they were attached to the latter; this in fact is the rigid layer. Further on, a thicker, more diffuse layer is formed mainly by positive ions, subject only to a weak attraction. This is called the "diffuse layer". Still, further on, one encounters an equal number of both positive and negative ions. This is the liquid medium free of any interaction with the particle.

The theory of the diffuse layer is based on Maxwell-Boltzmann and Poisson's fundamental equations. The electric potential, at a point of the solution, is defined in relation to geometric values and ionic concentrations.

These equations have often been revised and the successive improvements made to the theory have complicated the analysis introducing parameters having a difficult experimental approach.

That is why the approach of Gouy-Chapman's fundamental data can be considered as a basis, similar to the adjustments suggested by Stern, which have led to the notion of the Stern potential (see Figure 3).

The double layer consists of a single layer of fixed counterions (Stern layer) adsorbed to the charged surface and a diffuse layer of counterions distributed near the interface. Roughly 60 to 85% of the counterions are located in the Stern layer and are responsible for a sharp drop in potential across this layer (about 2 Å in thickness).

Therefore, on the surface there lies a layer of thickness in which the electric potential varies linearly from Ψ_0 to a Ψ_δ value known as the Stern potential.

The diffuse layer (Gouy-Chapman layer) has a thickness of up to several hundred Ångstrom and the potential gradually falls with increasing distance from the surface.

Electrokinetic effects appear with the modification of static balance conditions of a charged surface in a solution. Electrokinetic phenomena have been discussed by Sennet and Oliver.[2]

The theory of these effects is based on the concept of the zeta or electrokinetic potential. This potential is defined as being the value reached by the electric potential at a certain distance from the particle on a plane of shear.

That means that, under the influence of an applied electric field, the particle, along with the Stern layer, can move away from the diffuse ion layer at the plane of shear. The average potential in the surface of shear is called the electrokinetic or zeta potential.

Assuming this notion to be correct, there exists within the immediate vicinity of the particle a layer of liquid which moves along with it.

At the present stage of research, it can safely be said that the zeta potential is more representative of Ψ_δ than Ψ_o, and if it does not suffice for studying the diffuse layer, it can be used in a survey of the interactions taking place between different particles or between the surface of particles and electrolytes.

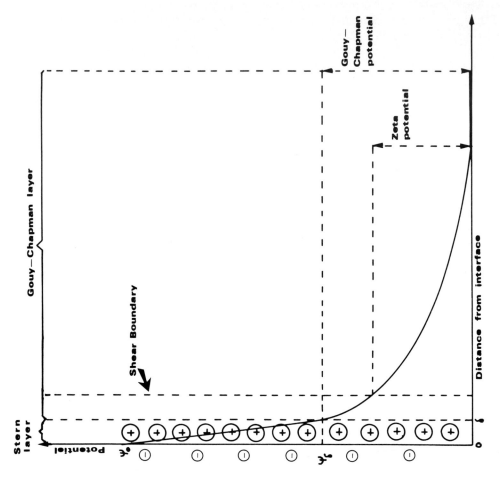

FIGURE 3.   Zeta potential.

Table 1 STREAMING POTENTIAL CHARACTERISTICS		**Table 2** SEDIMENTATION POTENTIAL CHARACTERISTICS	
Mobile phase	Liquid	Mobile phase	Liquid
Property measured	Potential	Property measured	Potential
Name	Streaming potential	Name	Sedimentation potential

IV. MEASUREMENT OF ZETA POTENTIAL

Methods for determining the zeta potential are generally based on the application of a potential or a pressure gradient.[5] In the case of a perturbing mechanical force, such as a difference in pressure at both ends of a liquid capillary column limited by charged surfaces, an electric potential called the streaming potential occurs. The inverse effect, obtained when charged particles are made to move by sedimentation in a still liquid, causes the appearance of a sedimentation potential.

When the perturbing force is of an electrical origin, the application of an electric field to both ends of a liquid capillary column limited by charged surfaces causes the liquid to flow. This is referred to as "electro-osmosis".

If an electric field is applied to a medium containing solid charged particles, the latter are activated in a certain direction, so that the particles of a given sign move towards the electrode of the opposite sign. This is "electrophoresis".

A. Streaming Potential (Table 1)

If a colloid suspension is forced through a capillary on a porous plug under a pressure gradient, the excess charges near the wall are carried along by the liquid and their accumulation downstream causes the buildup of an electric field which drives an electric current back against the direction of the liquid flow.

A steady state is quickly established and the measured potential difference across the capillary is called the streaming potential (E_s). E_s is related to the applied pressure difference (P) and the zeta potential.

B. Sedimentation Potential (Table 2)

If charged colloidal particles are allowed to settle under gravity, a sedimentation potential is generated. According to this property, the estimation of the zeta potential requires only a measurement of the potential developed between two points of different heights in the column. However, the method is not often used because it seems to be unreliable for general use.[6]

C. Electro-Osmosis (Table 3)

Instead of causing liquid to move through a porous plug (streaming potential), one can apply an electric field to do it. This is "electro-osmosis". The applied field acts upon the charges in the liquid and, as they move in response to the field, they drag the liquid along with them. Measurement of the current flow, the conductivity of the liquid, and the volume of the liquid transported gives information about the electric potential:

$$\int = \frac{4\pi\, Ve\, \eta\, \lambda}{D\, i}$$

where Ve is the electro-osmosis flow, η is the viscosity, λ is the specific conductance, D is the dielectric constant of the medium, and i is the current passing through the capillary or the porous plug.

Table 3 ELECTRO-OSMOSIS CHARACTERISTICS		Table 4 ELECTROPHORESIS CHARACTERISTICS	
Mobile phase	Liquid or none	Mobile phase	Particles
Property measured	Flow or pressure	Property measured	Particle mobility
Name	Electroosmosis	Name	Electrophoresis

When the electro-osmosis is analyzed in terms of the counterpressure required to obtain zero rate flow through the porous plug, the zeta potential may be calculated:

$$\int = \frac{4\pi \, Pe \, K}{D \, E}$$

where Pe is the electro-osmosis counterpressure, K is a constant, and E is the applied potential.

D. Electrophoresis (Table 4)

Electrophoresis is by far the most common procedure for determining the zeta potential. A large number of different techniques have been developed and different apparatus are commercially available, e.g., the electrophoretic mass transport analyzer, the zeta meter, the laser zee meter, and the Malvern zetasizer. If particles are induced to move by applying an electric field, measurement of their velocity provides information about their net electric charge. The motion of the particles can be followed in different ways: microscopic observation, direct weighing, etc. The zeta meter and the laser zee meter are based on microscopic observation while the electrophoretic mass transport analyzer is based on direct weighing of migrating particles. Both of these apparatus are used to measure electrophoretic mobility in dilute suspensions. If the zeta meter or laser zee meter are used, electrophoretic mobility is determined by particle velocity:

$$\mu = \frac{\dfrac{X}{t}}{\dfrac{V}{L}}$$

with X = distance covered, T = time, V = potential difference, and L = distance between electrodes.

Electrophoretic mobility is related to zeta potential by Smoluschonski's equation:

$$\mu = \frac{D \, \xi}{4\pi \, \eta}$$

where ξ = zeta potential, η = viscosity of liquid, and D = dielectric constant of liquid.

Several workers do not use zeta potential but only electrophoretic mobility to investigate electric double layer.[7-8]

Indeed, on the whole, the values of zeta potential are calculated with the values for η and D being constant and equal to those of distilled water. Therefore, zeta potential and electrophoretic mobility are linked by a constant factor and the values of electrophoretic mobility show more exactitude than those of zeta potential. Moreover, the conversion of electrophoretic mobility to zeta potential by Smoluschonski's equation is strictly only applicable to impenetrable, spherical particles.[9]

Table 5
INFLUENCE OF THE ZETA POTENTIAL UPON DISPERSION CHARACTERISTICS

Dispersion characteristics	Zeta potential		(mV)
Strong agglomeration and precipitation	+ 5		− 5
Threshold of agglomeration	− 5		− 15
Delicate disperison	− 16	to	− 30
Moderate stability	− 31		− 40
Fairly good stability	− 41		− 60
Very good stability			> − 61

V. MEAN APPLICATIONS OF ZETA POTENTIAL

The stability of a dispersed system is dependent upon zeta potential (Table 5). Even if electropositive values of zeta potential are possible, they are seldom found in nature. Most of the colloids, organic and inorganic, are electronegative when suspended in distilled water.

The degree of agglomeration or, on the the other hand, the stability of a colloid system, can be improved by good adaptation of the zeta potential.

It is adequate to classify the interactive forces involved between particles into three main classes.[9]

The first, long-range interactions, can result from van der Waals forces, which are usually attractive, and double-layer (electrostatic) interactions, which are repulsive. The second are short-range forces, which have been emphasized by Pethica.[10] They result from chemical bonds, e.g., electrostatic, covalent, and hydrogen bonds, dipole interactions, hydrophobic bonding, etc.[11] Most of the above forces act over short distances, typically less than 4 Å. They may be attractive or repulsive depending on the nature of the surface involved. Another class of short-range forces which has been discussed more recently is the result of the orientation of ordered liquid films at the solid-liquid interface.[12] Finally, interfacial reaction may take place at and around the contact area. One of the earliest theories of particle interaction associates the van der Waals attractive force with a double-layer repulsive force. It is the well-known DLVO (Derjaguin, Landau, Verwez, and Overbeek) theory. If one attempts to apply the DLVO theory, it is necessary to know the value of the surface potentials to calculate the electrostatic repulsion. These parameters cannot be as easily measured as the zeta potential or electrophoretic mobility.

VI. STUDY OF THE CELL SURFACE CHARGE PROPERTIES

A. Measurements of the Zeta Potential of Cells

The electrophoretic mobility is the best technique for measuring the zeta potential of microorganisms (bacteria and yeasts). For example, we used the laser zee meter apparatus to measure the zeta potential of yeast cells and it is particularly easy because cell dimensions are <100 μm and because their specific weight is not far from that of water.

These properties are very important for good, quick, and reproducible measurements. The laser zee meter model 500 provides means for ensuring that measurements are made at the stationary layer.[13]

Generally, before mobility measurements, the cells are washed with distilled water. The number of cells is limited to around 20×10^6 cells per milliliter; the voltage between the electrodes is adjusted between 50 and 150 V. The pH of the suspension is adjusted, but buffers are not used to fix the pH because ions could adsorb especially at the cell wall and modify the charge properties of the surface. The temperature is also noted. The measurements

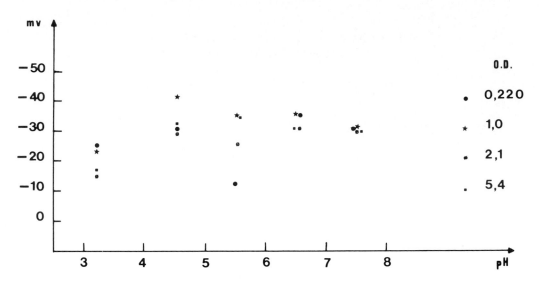

FIGURE 4.   pH influence on zeta potential values of haploid cells during growth in distilled water. pH is adjusted with HCl or NaOH. Absorbance of the culture; ● 0.220; ★ 1.0; ø 2.1; ■ 5.4.[13]

are performed as quickly as possible to avoid pH change caused by cellular material. The results are expressed in millivolts (zeta potential) or in microns squared per second per volt (electrophoretic mobility) as described before. Moreover, as shown recently for yeast, potential (Ψ) is linearly related to surface potential as described previously.[14]

B. Zeta Potential of Cells

A number of authors have studied the electrophoretic mobility of cells of microorganisms, but we are more specifically concerned with *Saccharomyces cerevisiae* yeast owing to its importance in the food industry and because it acts as a model in the fields of biochemistry and microbiology. The chemical composition of the wall of this organism is relatively well known; many workers have observed two regions containing an external layer consisting of phosphomamann proteins and an internal layer, formed mainly of glucans.[15-17]

Thus, the cell wall of yeasts, like that of plants, is composed almost entirely of polysaccharides: mannans and glucans.[17] The phosphorus is associated with mannans in the form of phosphodiesters and peptides to form phosphomannan proteins. Both the structure of peptides and their composition in amino acids are still unknown.

As regards the charge of this wall, Figure 4 illustrates the influence of pH on the zeta potential of the haploid cells of yeast in distilled water.[13] For pH values between 3 and 7, the potential ranges from -15 to -35 mV and the isoelectric point of this yeast strain is situated at values lower than 3. Van Haecht et al. have carried out the measurement of the N and P concentrations in the surface region of yeast cell walls in an original way using X-ray photoelectron spectroscopy (XPS or ESCA).[7] They confirm a correlation between the ratio N:P and isoelectric point of the cells. This is easily interpreted for, when the pH of the cell suspension is brought below 4, the ionogenic groups due to proteins are protonated and the net charge of the cell surface is mainly due to phosphate and amino groups; amino groups tend to make the surface more positive by holding a proton, while phosphate ions tend to make the surface more negative. Absolute measurements confirm the direct correlation with the N content and the inverse correlation with the P content.

Moreover, the change of culture conditions affects the P and/or N content. It is well-known that the intensity of aeration brings about changes in metabolic activity.[18]

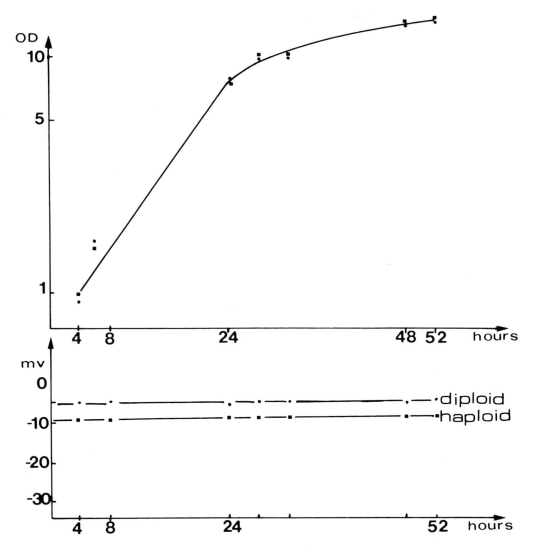

FIGURE 5. Zeta potential values as a function of growth for (●) diploid cells and (■) haploid cells.[13]

VII. STUDY OF THE MODIFICATION OF THE ZETA POTENTIAL OF MICROORGANISMS

A. Influence of Physiological Conditions on the Zeta Potential

Different phases can be observed during the growth of a microorganism (lag phase, log phase, and stationary phase). The exchanges taking place between the cell and the medium differ greatly during growth. It would be interesting to study electrophoretic mobility during an entire growth cycle. Using yeast as an organism, Thonart et al. have shown that the zeta potential of a cell did not vary during the growth cycle[13] (see Figure 5).

Identical results have been cited by Atkinson and Daoud[8] using bacteria, while Rouxhet has studied the growth of *Moniliella tomentosa*.[16] Figure 6 presents electrophoretic mobility data obtained for cells cultured in a cheap fermentor and harvested at different times during growth or during the resting phase. No significant difference associated with the age of the cells was observed.

However, the growth of a microorganism is often associated with production of a metabolite, with the consumption of the substrate, and thus leads to the modification of the

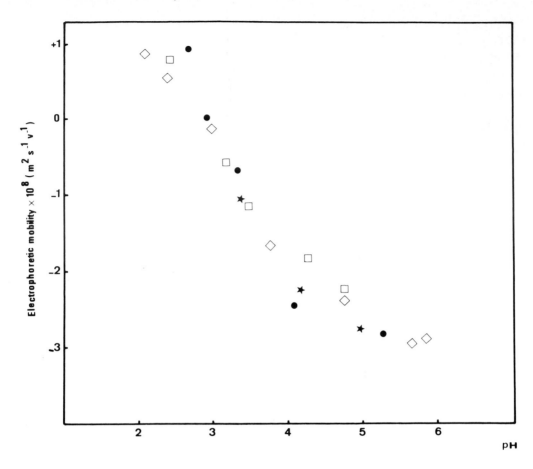

FIGURE 6.   Variation of the electrophoretic mobility of *Moniliella tomentosa* as a function of pH.[16,41] Cells grown in the Chemap fermentor and harvested after 43 hr (◇), 63 hr (□), 159 hr (★), 195 hr, i.e., in the resting phase (●).

cell environment. *Xanthomonas campestris,* for instance, produces an anionic polymer known as xanthan. According to culture conditions and growth phase, there is a great variation in the quantity of polymers, and we have established the influence of this polymer on the zeta potential of *Xanthomonas* cells (see Figure 7). The excretion of proteins and metabolites plays an important role in the charge of the wall.[19]

B. Modification of the Zeta Potential of Cells by the Addition of Polymers

The addition of anionic or cationic polymers modifies the zeta potential of cells. The influence of organic polymers such as proteins, inulin, and carrageenan has been studied on yeast cells. With cationic polymer (cationic gelatin and cationic starch)[20] it is possible to confer a net positive charge to the cells at pH 6 whereas with anionic polymers the charge evolves towards slightly more negative values (see Tables 6 to 8). These polymers are adsorbed on the ionized surface; this is probably a rather simplistic view as the cell surface is complex and rarely allows various types of interactions such as hydrogen bonds, electrostatic forces, etc.

However, this possibility of modifying the zeta potential by polymers will find its use in various biotechnological applications such as cell immobilization and flocculation and can be used to enhance transfer mechanisms between culture medium and the cell. In our opinion, these polymers offer the advantage of being neither toxic for the cell nor forbidden in food applications.

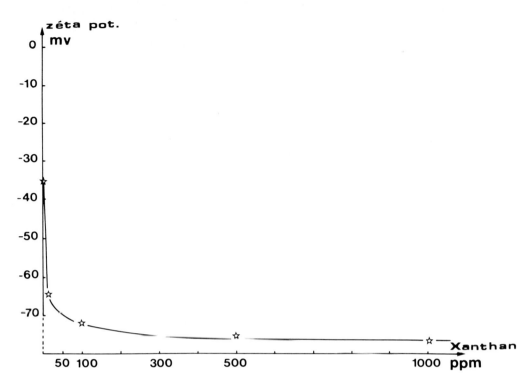

FIGURE 7.   Influence of xanthan addition on *Xanthomonas* zeta potential.[19]

Table 6
INFLUENCE OF VARIOUS AGENTS ON ZETA
POTENTIAL (mV) OF *SACCHAROMYCES*
***CEREVISIAE* IN DISTILLED WATER**

μg/mℓ	Inulin	Gelatin	Cationic starch	Proteose peptone	Carrageenan
0	−43	−40	−50	−36	−46
20	−32	+4	−10	−27	—
100	−38	+14	+6	−23	−42
200	—	—	+20	—	—
500	−35	+16	+23	−17	−34
1000	−29	+17	—	−12	−31
5000	—	—	—	−7	−26

C. Influence of Ions on the Zeta Potential of Cells

We have previously described how the ionic force of the medium influences the zeta potential of particles, essentially by reducing the double layer. What is more, it is worth mentioning again that the conversion of electrophoretic mobilities to the zeta potential uses the usual classical equations. It is implicitly assumed that the particle has an impenetrable surface. The problem of penetrable cell surface has been considered such as those of yeast and bacteria: at low ionic strengths the concept of impenetrability of the cell surface tends to be valid.[21]

Table 9 shows how the addition of various salts influences the zeta potential of yeast cells in different mediums at pH 5. According to the type of ion and its valency, the zeta potential varied in different ways. Small differences were observed with the addition of 1% NaCl, whereas 1% $CaCl_2$ or $Al_2(SO_4)_3$ led to a zero potential value. The effect of bivalent ions on yeast flocculation has been previously described and reviewed.[8-13]

Table 7
INFLUENCE OF VARIOUS AGENTS ON ZETA POTENTIAL (mV) OF *CANDIDA PSEUDOTROPICALIS* IN DISTILLED WATER

μg/mℓ	Cationic starch	Gelatin	Proteose peptone	Inulin	Carrageenan
0	− 35	− 35	− 35	− 35	− 35
2.5	− 20	—	—	—	—
5	+ 7	− 25	− 32	− 42	− 40
10	+ 11	− 5	− 31	− 45	− 46
20	+ 18	+ 1	− 31	− 48	− 47
50	+ 17	+ 8	− 30	− 52	− 48
100	+ 21	+ 15	− 29	− 53	− 49
500	+ 30	+ 20	− 23	− 53	− 51

Table 8
INFLUENCE OF VARIOUS POLYMERS ON ZETA POTENTIAL VALUES OF *KLUYVEROMYCES FRAGILIS* MEASURED ON DISTILLED WATER

μg/mℓ	Cationic starch	Gelatin	Inulin	Proteose peptone	Carrageenan
0	− 26	− 26	− 26	− 26	− 26
5	− 19	0	− 28	− 25	− 26
10	− 1	+ 3	− 30	− 25	− 31
20	+ 3	+ 5	− 32	− 25	− 35
50	+ 10	+ 11	− 35	− 24	− 39
100	+ 14	+ 16	− 37	− 23	− 40
500	+ 27	+ 18	− 37	− 20	− 42

Table 9
INFLUENCE OF VARIOUS AGENTS ON THE ZETA POTENTIAL OF YEAST CELLS IN MINIMAL MEDIUM (A) AND IN DISTILLED WATER (B) FOR HAPLOID AND DIPLOID

(%)	$CaCl_2$ (mV) A	$CaCl_2$ (mV) B	NaCl (mV) A	$Al_2(SO_3)_3$ (mV) (A)
0	− 7.8	− 39.3	− 7.8	− 7.8
0.01	− 6.0	− 26.4	—	− 5.8
0.05	− 3.5	− 18.3	—	—
0.1	− 1.5	− 11.3	—	− 4.0
0.2	—	—	—	0
0.5	− 0.5	− 6.4	—	0
1	− 0.1	± 0	7.0	—

It is known that the calcium atom can form a bridge between receptor sites, possibly carboxyl groups on different yeast cells. Moreover, the effect of ions on the zeta potential of cells depends on the medium; we have previously described the influence of this parameter. However, a competition can arise among the ions. In flocculation studies, for example, the NaCl antagonizes the action of the $CaCl_2$; the KCl has little effect. A number of ions, e.g., ferric, tin, and silver, have the ability to cause flocculation of a number of cells which are unaffected by the presence of calcium.[8]

Experiments carried out on the transport of ions can be extremely instructive if accompanied by studies of the wall charge. Borst-Pauwels and Theuvenet have reported that Ca^{2+} is more strongly bound to the negative groups on the cell membrane than Sr^{2+} in yeast.[22] An enhancement of divalent cation uptake by preloading the cells with phosphate would be due to the concomitant increase in the negative surface potential. Moreover, the zeta potential of the yeast cells appears to be much lower than the electrostatic potential experienced by the divalent cations in their translocation across the membrane. This is possible if the negative charges are not uniformly distributed over the surface of the membrane and if the transport of the cations occurs in areas of the membrane that are in close proximity to these negatively charged groups.

VIII. APPLICATIONS OF THE ZETA POTENTIAL IN BIOTECHNOLOGY

A. Improving the Yield of Immobilized Cells

The immobilization of cells consists of immobilizing living or dead cells in a reactor and thus making it possible to use them in a continuous process. Moreover, in immobilized cell reactors, the microorganisms must be retained at high cell concentration to achieve fast reaction rates. Various techniques facilitate the immobilization of cells in a reactor;[23-28] these can be grouped into four classes: immobilizaiton by adsorption, by covalent bond, by gel inclusion, and by flocculation without a support (cross-linking). Among these techniques, the one which interests us particularly is the immobilization by adsorption on a carrier.

Guidelines for prediction of adhesion are provided by the DLVO theory (as described previously), accounting for the long-range particle-particle interaction. The total interaction comprises two additive terms: one due to van der Waals forces and one due to electrostatic forces. However, it is recognized that the balance of short distance interactions (hydrogen bonds, dipole interactions) and van der Waals interactions at the particle-liquid-particle interfaces plays an important role.

Under physiological conditions (pH and temperature), the surface of microorganisms is negatively charged; if the support is also negatively charged, the cell-support interactions involve electrostatic repulsion terms which are unfavorable to adhesion.

To improve cell immobilization on support, it is necessary to decrease cell-support electrostatic repulsions. The value of the surface potentials must be known in order to calculate the electrostatic repulsion. These parameters cannot be easily measured. Electrophoresis or microelectrophoresis measures the potential shear plane, i.e., the zeta potential. In many cases (especially with microorganisms), it is assumed that the surface potential is equal to the zeta potential. Further on in this paragraph, we used zeta potential characteristics to optimize cell adsorption on different carriers. Optimization of the interaction between cell and support requires modification of the zeta potential either of the cell or the support. Particles having an opposite or nearly neutral charge are under favorable conditions. that cause them to become attached to one another.

Different proceedings have been used in the modification of the zeta protential of the cell or the support, based on the principles described previously. We have developed a technique of optimization.

One must note that cheap and readily available supports are generally negatively charged. (For example, glass beads, sawdust, etc.) A survey of the supports used in immobilizing cells by adsorption is given in Table 10. Studies dealing with measurement and the influence of various agents such as $CaCl_2$, NaCl, carboxymethylcellulose, $(Al)_2(SO_4)_3$, cationic starch, and cationic proteins have been described previously.[13-20]

If you add a cationic polymer to a cell suspension and to sawdust (e.g., cationic starch

Table 10
SUPPORTS USED IN
IMMOBILIZING CELLS BY
ADSORPTION

Support	Microorganisms
Wood particles	*Saccharomyces*[20]
Ion-exchange resin	*Saccharomyces*[29]
Polystyrene	*Pseudomonas*[30]
Sawdust	*Saccharomyces*[13]
Ceramic	*Saccharomyces*[29]
Silica beads	*Saccharomyces*[31]
Glass beads	*Acetobacter*[32]
Dualite AIOID (acetate)	*Saccharomyces carlsbergensis*[33]
Spherosil XOB 015	*Saccharomyces carlsbergenesis*[33]

or cationic gelatin), the zeta potential of cells becomes positive while that of sawdust remains negative (see Figures 8 and 9 and Table 11). This is not the case with inulin; e.g., the zeta potential of cells and sawdust remains negative irrespective of the amount of inulin added. When cells are immobilized by passing a suspension of yeast on a column of sawdust, the attachment yield will be strongly influenced by the kind of polymers added and by the concentration of this polymer.

With *Saccharomyces cerevisiae* the yield of immobilized cells varied from 27 to 66 mg of cells per gram of support with addition of gelatin (0.2%), and to 68 mg/g with the addition of cationic starch (0.01%). No effect was noticed with inulin.[13-20] A correlation has been established between values of the zeta potential and cell immobilization on sawdust.

There are other techniques for modifying the zeta potential of cells. Starvation of yeast cells (*S. cerevisiae*) in pure water provolves their adhesion to glass or polycarbonate; the efficiency of the treatment was shown to be related to both a release of substances, decreasing the cell-cell and cell-support electrostatic repulsions and modification of the cell wall itself.[34-35] Adhesion of yeast to various carriers was also achieved by adding metallic-adsorbing metallic ions on the cells or the support or by coating the support with a layer of positively charged colloidal particles which act as a binding agent between the cells and the supports.

Several authors use pH conditions or the physiological state of cells for optimizing the immobilization on glass or polystyrene.[34] Rouxhet et al. have measured the degree of coverage (%) of the supports by *Moniliella tomentosa* at various pH values.[34] The increase in the amount of cells immobilized on glass, as the pH decreases from 5 to 3.5, is attributed to a decrease in the surface negative charge of both the cells and the supports. Cell immobilization is set up to improve the performance of a continuous fermentation process by increasing the amount of cells per reactor volume but the performances of a continuous reactor do not depend solely on the quantity of immobilized cells. Let us consider, for example, the production of alcohol. In a reactor, it is possible to immobilize in more or less 6 hr a quantity of cells varying between 100 and 200 mg of cells per gram of support, and to obtain a productivity rate situated between 20 and 40 g of ethanol per liter per hour using a concentration of ethanol of around 70 g/ℓ.

Certain authors have particularly focused their attention on the formation of a single layer of cells on the support, for this single layer adheres particularly well to the support, and cell retention according to the rate of flux is also especially favorable.[21] However, it is obvious that cellular concentration increases when the reactor is in operation causing the

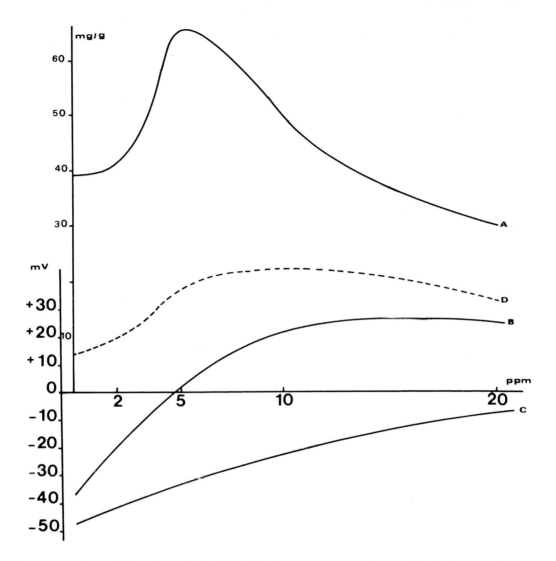

FIGURE 8. Influence of cationic starch on zeta potential of (B) *Candida* and (C) sawdust in distilled water. (D) = the difference between cells and sawdust zeta potential. (A) = yield of immobilization (mg cells/g carriers).[20]

buildup of a multiple layer stem. This type of reactor can operate for periods of up to 6 months and the main question which arises is the importance of the zeta potential during operating. It is clear that the zeta potential influences the quantity of immobilized cells during the 1-g phase; however, the physiological mechanisms (production of metabolites) and mechanical retention can play an important part during the operating of this kind of reactor.

Controlling the zeta potential of cells and the support is often of help in understanding cell desorption and in reducing the operating time of the reactor.

B. The Immobilization of Enzymes

The immobilization of enzymes on a solid support is a technique employed in order to retain and reuse these enzymes, which in turn leads to a decrease of the production costs. The properties of immobilized enzymes often differ appreciably from those of free enzymes, being either favorable or unfavorable. Stability in relation to pH or temperature are two such

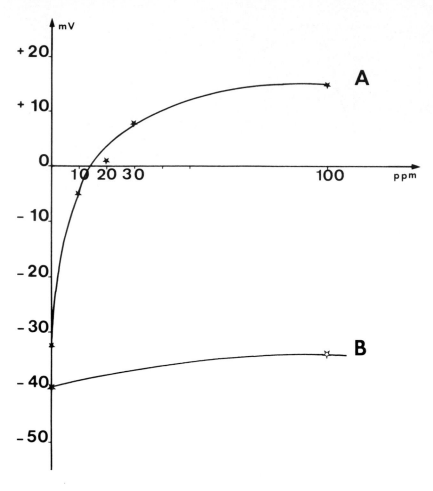

FIGURE 9.   Influence of gelatin on zeta potential of (A) *Candida* and (B) sawdust in distilled water.[13]

Table 11
INFLUENCE OF VARIOUS AGENTS ON ZETA
POTENTIAL (mV) OF *CANDIDA PSEUDOTROPICALIS* IN
DISTILLED WATER

$\mu g/m\ell$	Cationic starch	Gelatin	Proteose peptone	Inulin	Carrageenan
0	− 35	− 35	− 35	− 35	− 35
2.5	− 20	—	—	—	—
5	+ 7	− 25	− 32	− 42	− 40
10	+ 11	− 5	− 31	− 45	− 46
20	+ 18	+ 1	− 31	− 48	− 47
50	+ 17	+ 8	− 30	− 52	− 48
100	+ 21	+ 15	− 29	− 53	− 49
500	+ 30	+ 20	− 23	− 53	− 51

Table 12
INFLUENCE OF ACTIVATION AGENTS (DPMD AND TEA) ON
SUPPORT ZETA POTENTIAL IN FUNCTION OF pH

	pH						
AVICEL PH 101	**3.0**	**3.6**	**4.0**	**4.6**	**5.6**	**6.6**	**8.0**
Nonactivated	− 6.0	− 19.8	− 31.2	− 36.5	− 39.0	− 41.4	− 45.6
2.5% DPMD	+ 7.1	− 4.5	− 15.7	− 20.5	− 28.0	− 30.8	− 28.8
2.5% DPMD + 2.5% TEA	+ 46.1	+ 39.4	+ 30.7	+ 16.5	− 18.1	− 36.7	− 38.7
2.5% DMPD + 5.0% TEA	+ 50.3	+ 46.7	+ 43.1	+ 38.0	+ 18.7	− 10.5	− 21.0
2.5% TEA	− 7.1	− 14.5	− 20.5	− 28.5	− 32.1	− 29.8	− 27.0
2.5% TEA + 5.0% DPMD	+ 46.0	+ 43.5	+ 38.0	+ 25.0	− 14.0	− 18.4	− 26.4

examples. These differences in behavior are often ascribed to modifications of the microenvironment of enzymes on the support.[36] However, too little research has been conducted on the characteristics of this microenvironment, and therefore, the performance of an immobilized enzyme is often the result of empirical research.

The measurement of the zeta potential at the support-enzyme interface must be considered a tool in the study of the microenvironment of an immobilized enzyme, as illustrated by the following example concerning the immobilization of β-galactosidase of *Aspergillus niger* on the microcristalline cellulose AVICEL.[37] We shall study the influence of the carrier zeta potential after activation and also the zeta potential of immobilized enzymes.

1. Activation of the Support

It is therefore possible to activate a support while giving it different electrokinetic characteristics either by choosing the activation agents or by modifying their respective proportions (Table 12). When a support (microcristalline cellulose AVICEL PH 101) is activated by reaction with diphenylmethane 4-4′ diisocyanate (DPMD) and triethylamine (TEA), its zeta potential is modified and rendered positive to acid pH and in particular to the pH (pH 3.6) used during the enzyme immobilization (Table 12).

2. Immobilization of β-Galactosidase

There is a noticeable decrease in immobilization yield and residual activity due to an increase in the concentration of β-galactosidase (Table 13). Moreover, the zeta potential of the activated support changes with the addition of β-galactosidase. For example, at pH 3.6, the evolution follows a decrease of the zeta potential when β-galactosidase is added (see Figure 10 and Table 13). The isoelectric point of β-galactosidase of *Aspergillus niger* is close to 4.6.[38] Below this pH, the net charge of the protein as well as the activated support are both positive, although the zeta potential resulting from their interaction progresses toward less positive values. It would seem from this behavior that negative charges of the enzyme play a role in the reaction between the enzyme and the support. These negative charges probably originate from aspartic and glutamic acid which represent more than 20% of the composition in amino acids of β-galactosidase.[38]

3. Influence of the Substrate

When a support is charged, interactions can take place within the medium and also with the substrate if it is also charged. A positively charged support will therefore attract or repulse the substrate according to whether the latter is negatively or positively charged. In

Table 13

INFLUENCE OF β-GALACTOSIDASE CONCENTRATION ON CHARACTERISTICS OF IMMOBILIZATION (IMMOBILIZATION YIELD, RESIDUAL ACTIVITY, ZETA POTENTIAL)

Enzymes concentration (mg/g support)	Immobilization yield	Residual activity	Zeta potential[a] (at pH 3)
0	—	—	39.4
15	99.6	79.5	20.6
100	93.6	43.0	14.2
250	39.8	27.5	12.8
500	28.5	21.0	2.0
1250	17.5	3.7	0

[a]   Prior to immobilization.

both cases, the concentration of substrate surrounding the complex enzyme support will differ from that of the medium, often implying the modification of Michaelis Menten's apparent constant of the immobilized enzyme.

This constant increases if both charges are of the same sign, and decreases if they are of opposite signs.[39] It is possible to link the substrate concentration (S_s) surrounding the support and the substrate concentration (S_m) in the solution mass. This relation calls on the value of the electrostatic potential (Ψ):

$$S_s = S_m \exp \left(- \frac{Z E \psi}{K T} \right)$$

where Z = coefficient, E = electron charge (absolute value), K = Bolztmann's constant, and T = absolute temperature. By combining this equation with that of Michaelis and Menten, we obtain a relation linking Michaelis-Menten's apparent constant (K'_m) of immobilized enzyme with that of the enzyme in solution (K_m). This relation involves the electrostatic interactions via the electrostatic potential:

$$K'_m = K_m \exp \left(- \frac{Z E \psi}{K T} \right)$$

Assuming that the support and the substrate are of the same sign, Z and Ψ are of different signs and the apparent coefficient of Michaelis-Menten of the enzyme support complex is superior to that of the enzyme in solution. In consequence, the measurement of the zeta potential is of particular interest in this field of application, as it provides a means for measuring Ψ. Nevertheless, all the substrates for the enzymes are not necessarily charged. Lactose, for instance, a substrate of β-galactosidase, belongs to this category. Even in this situation, the knowledge of electrokinetic potential and its evolution can present a certain interest. Two major effects appear when β-galactosidase is fixed in the presence of lactose.[37]

On the one hand, lactose causes a decrease in the adhesion yield. On the other hand, it provokes an increase in residual activity of the fixed enzyme. This beneficial effect comes as no surprise for it is generally recognized that the immobilization of an enzyme in the presence of its substrate enhances residual activity. Indeed, the substrate plays a protective role in relation to the active site of the enzyme. However, there is competition between the enzyme and its substrate for the immobilization sites. Paquot and Hasnaoui have demonstrated

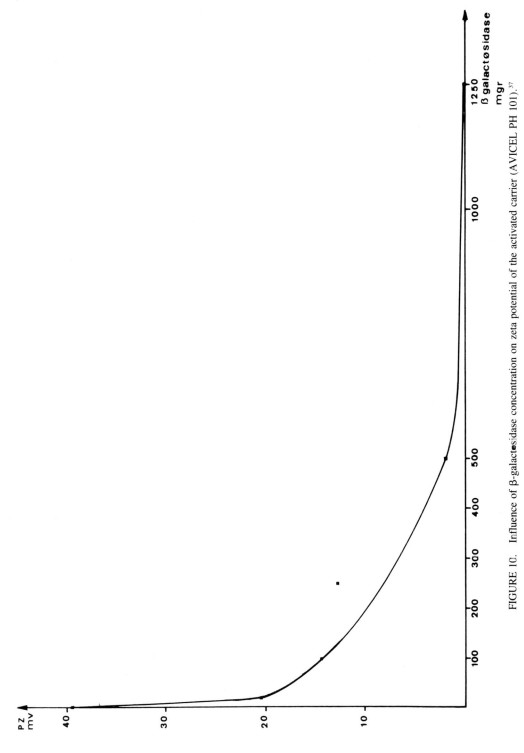

FIGURE 10. Influence of β-galactosidase concentration on zeta potential of the activated carrier (AVICEL PH 101).[37]

how more than 100 mg of lactose could adhere to 1 g of activated carrier. Zeta potential evolution is in relation to the competition between lactose and β-galactosidase.

4. Optimum pH of Immobilized Enzyme

The immobilization of an enzyme entails the modification of its microenvironment compared to that of an enzyme in solution. The pH with regard to the interface differs from that of the solution. A negatively charged support has a tendency to attract protons with the ensuing decrease of the pH surrounding the immobilized enzyme. The opposite situation can be observed when the support is positively charged.

The pH at the interface (pH_s) is linked to the pH of the solution (pH_a) by a relation where electrophoretic mobility is present. This equation was established by Hartlez and Roe:[40]

$$pH_s = pH_a + 0.325\mu$$

We should, however, bear in mind the fact that when working in an aqueous medium similar to that of distilled water (viscosity 1 cpoise, dielectric constant: 80.1), mobility can be converted to the zeta potential by multiplying by 14.1.

This also means that the readout of -56.4 mV for the zeta potential corresponds to a mobility of -4.0 cm^{-1} V^{-1} sec^{-1} for μ. In these conditions:

$$pH_s : pH_a - 1.3$$

Expressed in other terms, from the equation of Hartlez and Roe it would appear that the pH around the support medium can differ from that of the medium by at least one unit.

The apparent optimum pH of the immobilized enzyme can consequently be modified and this effect can be attributed to the electrostatic nature of the support. Paquot and Hasnaoui have illustrated this with the immobilization of β-galactosidase.

C. Zeta Plus Filters

We will not go into detail about this application of the zeta potential in biotechnology, but will simply describe its essential characteristics. The filtering medium zeta plus combines filtration by sieving and by electrokinetic absorption. The greater part of colloids and microorganisms have a negative electrokinetic potential and can be absorbed on the filter zeta plus. The biggest particles are captured by sieving and the other by electrokinetism.

IX. CONCLUSION

The zeta potential may be a good tool for improving different biotechnological processes. Indeed, zeta potential measurements give some information about interfacial properties. Different techniques could be used for zeta potential measurement. Microelectrophoresis seems to be very useful for measurement of the zeta potential of cells. This parameter is important for improving cell retention on a carrier. Moreover, a continuous reactor with immobilized enzymes can be optimized by adaptation of the interfacial properties.

The zeta potential should also be an important parameter in explaining the mass transfer phenomena between the medium and the cell.

ACKNOWLEDGMENTS

We thank Mrs. Gregoire and Mrs. Bock for the excellent technical assistance.

REFERENCES

1. **Berkeley, R. C. W., Lynch, J. M., Melling, J., Rutter, P. R., and Vincent, B.,** Microbial Adhesion to Surfaces, Ellis Harwood, London, 1980.
2. **Sennet, P. and Oliver, J. P.,** Colloidal dispersions. Electrokinetic effect and the concept of zeta potential, *Ind. Eng. Chem.,* 57, 32, 1965.
3. **Ross, S. R. and Long, R. F.,** Electrophoresis as method of investigating electric double layer, *Ind. Eng. Chem.,* 10, 58, 1969.
4. **Riddick, T. M.,** Zeta potential and Polymers, *J. Am. Water Work Assoc.,* 58, 6, 719, 1966.
5. **Paquot, M., Thonart, Ph., Flambert, C., Deroanne, C., Fraipont, L., Coppens, R., and Mottet, A.,** Le Potentiel zéta: Possibilité d' application en Industries papetière techniques de mesure, *Ann. Gembloux,* 83, 253, 1977.
6. **Hunter, R. J.,** *Zeta Potential in Colloid Science. Principles and Applications,* Academic Press, London, 1981.
7. **Van Haecht, J. L., Defosse, C., Van den Bogaert, B., and Rouxhet, G.,** Surface properties of yeast cells: chemical composition by XPS and isoelectric point, *Colloids Surf.,* 4, 343, 1982.
8. **Atkinson, B. and Daoud, J. S.,** Microbial flocs and flocculation, *Adv. Biochem. Eng.,* 4, 42, 1976.
9. **Tadros, Th. F.,** Particles-surface adhesion, in *Microbial Adhesion to Surfaces,* Ellis Horwood, Chichester, 1980, 93.
10. **Pethica, B. A.,** The physical chemistry of cell adhesion, *Exp. Cell Res. Suppl.,* 8, 123, 1961.
11. **Kauzmann, W.,** Some factors in the interpretation of protein denaturation, *Adv. Protein Chem.,* 14, 1, 1959.
12. **Pashley, R. M. and Kitchener, J. R.,** Surface forces in adsorbed multilayers of water on quartz, *J. Colloid Interf. Sci.,* 71, 491, 1979.
13. **Thonart, Ph., Custinne, M., and Paquot, M.,** Zeta potential of yeast cells: application in cell immobilization, *Enzyme Microb. Technol.,* 4, 191, 1982.
14. **Theuvenet, A. P. R. and Borst-Pauwels, G. W. F. H.,** Effect of surface potential on Rb^+ uptake in yeast, *Biochim. Biophy. Acta,* 734, 62, 1983.
15. **Mozes, N. and Rouxhet, P. G.,** Deshydrogenation of cortisol by *Arthrobacter simplex* immobilized as supported monolayer, *Enz. Microb. Technol.,* 6, 497, 1984.
16. **Rouxhet, P.,** personal communication.
17. **Arnold, W. N.,** *Yeast Cell Envelopes: Biochemistry, Biophysics and Ultrastructure,* Vol. I, CRC Press, Boca Raton, Fla., 1981, chap. 5.
18. **Oura, E.,** Effect of aeration intensity on the biochemical composition of baker's yeast, *Biotechnol. Bioeng.,* 16, 1197, 1974.
19. **Thonart, Ph., Paquot, M., Hermans, L., Alaoui Hammedi, and d'Hippolito, P.,** Xanthan production by *Xanthomonas campestris* NRRIB-1459 and interfacial approach by zeta potential measurement, *Enzyme Microb. Technol.,* 7, 235, 1985.
20. **Michaux, M., Paquot, M., Baijot, B., and Thonart, Ph.,** Continuous fermentation: improvement of cell immobilization by zeta potential measurement, *Biotechnol. Bioeng. Symp.,* 12, 475, 1982.
21. **Kayem, G. J. and Rouxhet, P. G.,** Adsorption of colloidal hydrous alumina on yeast cells, *J. Chem. Soc. Faraday Trans.,* 1, 561, 1983.
22. **Borst-Pauwels, G. W. and Thevenet, A. P. R.,** Apparent saturation kinetics of divalent cation uptake in yeast caused by reduction in the surface potential, *Biochem. Biophys. Acta,* 771, 171, 1984.
23. **Abbot, B. J.,** Immobilized cells, in *Annual Reports on Fermentation,* Perlman, D., Ed., Academic Press, New York, 1977, 1.
24. **Chibata, I. and Tosa, T.,** Transformation of organic compounds by immobilized microbial cells, *Adv. Appl. Microbiol.,* 22, 1, 1977.
25. **Jack, T. R. and Zajic, J. E.,** The immobilization of whole cells, *Adv. Biochem. Eng.,* 5, 125, 1977.
26. **Durand, G. and Navarro, J. M.,** Immobilized microbial cells, *Process Biochem.,* 13, 14, 1978.
27. **Lebesque, Y. and Dubreuil, P.,** Cellules immobilisées, *Bio-Sciences,* 2, 7, 107, 1983.
28. **Kennedy, J. F. and Cabral, J. M. S.,** *Immobilized Microbial Cells,* Vol. 4, Chibata, I. and Wingard, L. B., Jr., Eds., Academic Press, New York, 1983, 189.
29. **Daugulis, A. J., Brown, N. M., Cluett, N. R., and Dunlop, D. B.,** Production of ethanol by adsorbed yeast cells, *Biotechnol. Lett.,* 3, 11, 651, 1982.
30. **Fletcher, M.,** The effects of culture concentration and age time and temperature on bacterial attachment to polystyrene, *Can. J. Microbiol.,* 23, 1, 1977.
31. **Navarro, J. M. and Durand, G.,** Modification of yeast metabolism by immobilization onto porous glass, *Eur. J. Appl. Microbiol.,* 4, 243, 1977.
32. **Ghommidh, C., Navarro, J. M., and Durand, G.,** Acetic acid production by immobilized *Acetobacter* cells, *Biotechnol. Lett.,* 3, 93, 1981.

33. **Navarro, J. M.,** Immobilisation de *Saccharomyces uvarum* par adsorption sur des granulés de briques, *Sci. Aliments,* 1, 4, 513, 1981.

34. **Rouxhet, P. G., Mozes, N., Van Haecht, J. L., Reuliaux, L., and Palm-Gennen, M. H.,** Immobilization of microbial cells by adhesion to a support, *Third European Congress on Biotechnology,* Vol. 1, Verlag-Chemie, Munchen, 1984, 319.

35. **Van Haecht, J. L., De Bremacker, M., and Rouxhet, P. G.,** Immobilization of yeast by adhesion to a support without use of a chemical agent, *Enzyme Microb. Technol.,* 6, 221, 1984.

36. **Durand, G. and Monsan, P.,** Les enzymes immobilisées, *Séries de synthèses Bibliographiques C.D.I.U.P.A.,* APRIA Massy, 1974, 5.

37. **Paquot, M. and Hasnaoui, A.,** Aspects électrocinétiques lors de l'immobilisation de la β-galactosidase sur un support solide, *Lebensm. Wiss. U. Technol.,* 19, 17, 1986.

38. **Windner, F. and Leuba, J. L.,** β-Galactosidase from *Aspergillus niger* — separation and characterization of three multiple forms, *Eur. J. Biochem.,* 100, 553, 1979.

39. **Hornby, W. E., Lilly, M. D., and Crook, E. A.,** Some changes in the reactivity of enzymes resulting from their chemical attachment to water-insoluble derivative of cellulose, *Biochem. J.,* 107, 669, 1968.

40. **Mac Laren, A. D. and Pacher, L.,** Some aspects of enzyme reaction in heterogeneous system, *Adv. Enzymol.,* 33, 245, 1970.

41. **Rouxhet, G.,** personal communication.

Chapter 4

ENERGY CONSERVATION IN AEROBIC DIGESTORS BY BACTERIAL AUGMENTATION

Morton W. Reed

TABLE OF CONTENTS

I. INTRODUCTION

For a general discussion of energy conservation in wastewater treatment plants, the reader is referred to References 1 to 3. This study deals specifically with aerobic digestors and bacterial augmentation. It is based on the results of laboratory and full-scale testing conducted by the Tennessee Valley Authority (TVA) and the city of Columbia, Tenn.[4,5]

Any conventional activated sludge plant will generate solids which must be wasted. The design capacity of plants and the amount of capital available have tended to influence the choice of anaerobic vs. aerobic sludge digestors. A small plant of less than 10 million gallons per day (MGD) capacity will have less capital to spend for sludge digestion than say a 50 MGD plant that opts for anaerobic digestion at a higher first cost but with lower overall operating costs. From an energy perspective, the aerobic digestor requires much more power for operation than the anaerobic digestor, which produces methane gas. Gas holders, heat exchangers, and combustion control equipment make the anaerobic digestor cost more initially. A return on this investment is expected from the methane gas produced.

As electric power costs increase, the decision is made to allocate more capital to new plants so that anaerobic digestors can be installed. Retrofitting aerobic digestors to anaerobic ones is then attractive.

Another approach is to make aerobic digestors use less power for aeration. This would give activated sludge plants the best features of both a low capital cost and low operating costs.

II. BACTERIAL AUGMENTATION STUDIES AT A MUNICIPAL PLANT

The premise of bacterial augmentation is that conventional biological waste treatment plants owe their success to the persistence of soil bacteria in finding their way into the system.[6] There have been many case histories reported[7-11] where bacterial augmentation has improved removal efficiencies and eliminated odor problems, foaming, and poor dewaterability. Controlled laboratory studies are rarely included in the case histories.

At the time of the initial discussions (9/25/81) on aeration energy savings, the plant inlet flow was about 3.8 MGD. The aerated volume was as follows:

4 Digestors	201,604 ft^3
4 Aeration basins	336,052 ft^3
2 Pre-aeration basins	20,743 ft^3
4 Sludge holding tanks	37,506 ft^3
Total aerated volume	595,905 ft^3

The plant has three 300-hp centrifugal blowers. The amount of power consumed by the blowers fluctuates with dissolved oxygen concentrations, the total volume being aerated. Plant design calls for continuous aeration at about 24 to 30 cubic feet per minute (cfm) per 1000 ft^3. This requires running two blowers with a third on standby for rotation.

The plant has recently cut the number of aeration basins to two because inlet flows are only about 4 MGD. The number of aerobic digestors in service had remained at four. This has resulted in the continuous running of only one blower with occasional need for using a second blower to combat hot weather surges in oxygen transfer demand and odor problems.

Each blower has a minimum operating point to avoid surge conditions and a maximum operating point to prevent motor overload. The manufacturer recommends operating with a flow 10% above the surge point.

Figure 1 is the result of submetering and flow measurement work done by the TVA onsite. The plant can usually operate on 20 cfm per 1000 ft^3 and keep dissolved oxygen levels in

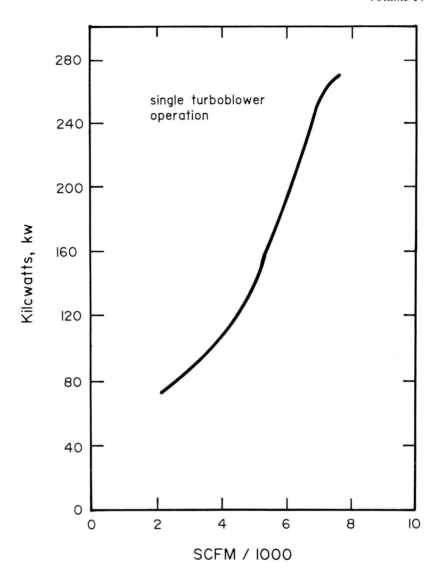

FIGURE 1. Turboblower performance curve.

the range 1 to 2 ppm. TVA laboratory results (see Section III) using bacterial additives and batch digestion cycles were sufficient to warrant full-scale testing. The amount of air needed in the lab study was reduced from 8 hr to as little as 2 hr per shift. Because it takes from 8 to 10 days to fill a digestor, the digestors were operated as semibatch units. A master test plan (Table 1) of several batch cycles was set up. Each cycle was monitored to optimize the amount of additive used and to evaluate overall performance. The plant operators were given the task of checking dissolved oxygen (DO) levels and cycling air and sludge values. The plant laboratory measured the solids levels, PH, supernatant, and dissolved nitrate in each digestor. A cycle consisted of continuous aeration during filling with cycling of aeration on the 3rd day through completion of a cycle. Total cycle time depends on the time used in filling the other digestor. The lab determined the amount and frequency of additive dosage. About 1 ppm/day was added on the first cycle. Because the plant inlet flow was less than design, the number of semibatch digestors was cut in to two. The test plan was used for about 2 months with no significant deterioration in digestor performance. Because of the

Table 1
MASTER TEST PLAN

Test day		Digester status		Digester status	
	(prev.)		Drained		Full-digesting — with flow
	1—8	(6—15)	Fill		Digesting — no flow
	9		Fill		Decant
	10—18		Digesting cycle 1	(6—25)	Fill
(7—4)	19		Decant		Fill
	20—27	(7—10)	Fil		Digesting cycle 1
	28		Fill		Decant
	29—37		Digesting cycle 2	(7—18)	Fill
	38		Decant		Fill
(7—24)	39—47		Fill		Digesting cycle 2
	48		Fill		Decant
	49—57		Digesting cycle 3		
	58		Decant		

Note: Actual flow rates will change number of days per cycle. Day 1 date was June 15, 1982. Fill time = 9 days; total batch cycle = 19 days; 1 day decant; digester 4 on standby; digester 1 for supernatant holding.

ability of the additive to control odors due to anaerobic conditions, a second blower was not used during the summer months. During digestor cycling, a kilowatt demand meter showed that aerating a single digestor requires about 13 kW. This is about 400 to 500 cfm per digestor which is at most 10 cfm per 1000 ft^3 of aerated volume. This is considerably less than 20 cfm/1000 ft^3. Increasing the flow of air to the digestors to about 20 cfm/1000 ft^3 did not improve the DO concentration.

A. Recommendations

The number of digestors being aerated was reduced from four to two. This, combined with the additive program, will allow the plant to avoid using two blowers during hot weather when odor problems were likely. From Figure 1, an estimated 184 kW is being saved by not operating a seasonal blower. Assuming that this is avoided for 2 months each year, the annual saving is

$$(184 \text{ kW})(\$7.14/\text{kW})(2 \text{ months}) = \$2,627.52$$

TVA laboratory and full-scale tests at the Columbia plant have shown that it is possible, when using commercially available liquid mixed culture bacterial additives (MCB), to reduce the amount of digestor aeration. The amount of aeration, the amount of additive, and the frequency of dosage depends on the digestor design and sludge properties of the individual plant. As a preliminary guide, the amount of MCB per dose for digestors is 1 to 2 ppm. The frequency of dosage depends on the flow scheme. For a batch operation, the frequency may be less than that for a continuous-flow system. Daily dosage is recommended for a continuous-flow system. The plant must make laboratory measurements if digestor perform-ance, including nitrate levels, to determine if the additive is effective. An important function of the additive should be to utilize nitrate during anaerobic conditions when aeration is interrupted. This probably prevents hydrogen sulfide formation from sulfates entering the digestors. According to lab studies, the MCB method will not work unless the mixed liquor is subjected to periods or zones of alternating aerobic-anaerobic conditions. This can be achieved in a continuous-flow system by providing zones where air diffusers are shut off and dispersion in minimized by baffles or partitions. Because of mixing considerations, this

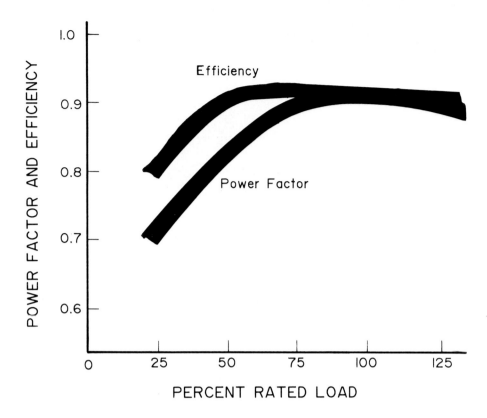

FIGURE 2.   Efficiency and power factor of three-phase motor (75 to 200 hp).

may not be practical. Attempting to use basins in series where air is completely cut off in individual basins will probably result in flow and mixing problems. If mechanical mixers are used in combination with separate aeration equipment, this may be practical. For plants that have a continuous-flow system that uses diffusers for aeration and mixing it is suggested that all the digestor basins have their aeration cycled on a sequential schedule.

Semibatch operation results in the digestors requiring about 8 days to fill. During a filling period aeration may be a maximum of 6 hr per shift (to allow for 2 hr of aeration needed for the next digestor). The energy conservation opportunity (ECO) calculations for the kilowatt hours are simplified if this is ignored. The error is not significant because the actual air flows will depend on the sludge flow rates. For the test period, two digestors were used on an alternating semibatch schedule. A third digestor was being used to provide capacity for excess sludge held up prior to dewatering. The procedure for cycling two digestors is as follows. At the beginning of each shift the operator would find full digestor basins with no air flow. He would open the air valve of one digestor to the operating point [about 500 standard cubic feet per minute (scfm)]. After about 2 hr of aeration, he would close the air valve on the first digestor and then open the air valve on the next digestor to its operating point. He would close this valve after about 2 hr. Throttling of the 300-hp blower exhaust may help if surge conditions develop. Throttling of blowers is preferred over frequent starting and stopping of the blower motors. Motor efficiency drops rapidly as the load drops below 50%. The power factor is even more sharply reduced as shown in Figure 2. An estimate is made based on the aeration schedule of Figure 3. The laboratory staff would have the responsibility for computing the amount of additive needed and adding it to the first digestor during aeration on the first shift. For example, if the volume in the digestor were 377,000 gal, then a 1-ppm concentration would require 0.377 gal (48 oz).

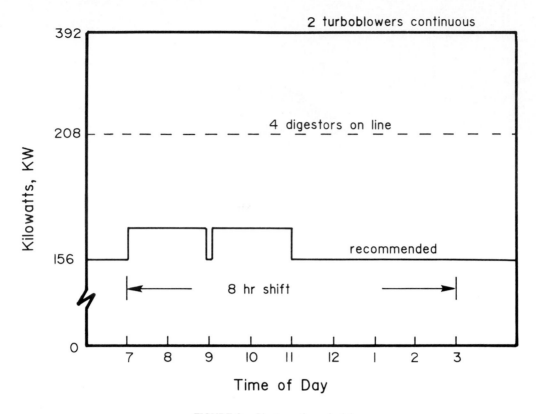

FIGURE 3.   Plant aeration schedule.

For a typical cycle the plant actually used 729 oz or 5.7 gal of additive during a 20-day period for one digestor. Using a typical cost of $14 per gallon, the annual cost for both digestors is

$$(5.7 \text{ gal})(365/20)(\$14)(2) = \$2,910$$

The amount of additive used is subject to refinement. Using Figure 4 as an illustration, the electrical energy demand is reduced by 26 kW. The annual savings is

The kWh reduction per shift is

$$26(4) + (26)(8) + 4(13) = 364 \text{ kWr/8 hr} + (26)(4)$$

Annual kWh saving is

$$\left(\frac{24}{8}\right)(365)(364) = 398,580 \text{ kWh} \times \$0.0264/\text{kWh} = \$10,522$$

Total savings = $12,750

An additional demand reduction of 13 kW would have been achieved if the third digestor had not been needed. This is not claimed in the ECO.

To further reduce the blower power, air flows less than that required to maintain 20 cfm/1000 ft^3 of liquid can be attempted. If DO levels stay above 1 ppm, then the air flow is

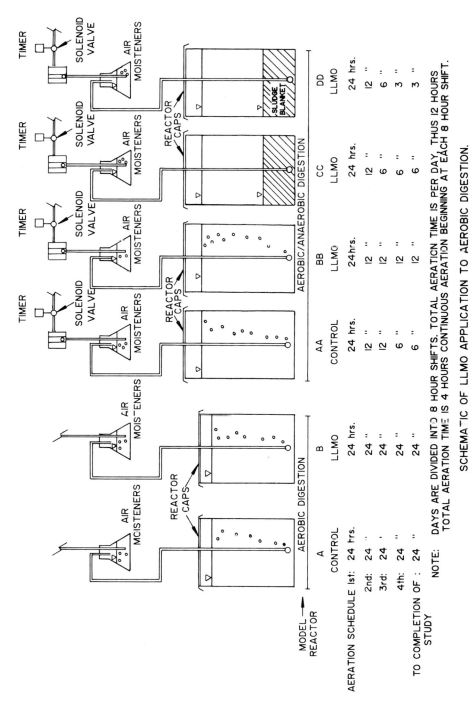

FIGURE 4. Experimental data.

probably sufficient. Clogging of diffusers is a potential problem that is present with or without cyclic aeration. Cycling diffusers off will aggravate a clogging or plugging condition. A trial period is needed to determine if cyclic operation can be implemented. Mixing may be reduced at lower air flow rates.

Some vendor names are listed below. Others may be found in the *Thomas Register* or a similar directory. The TVA does not endorse any particular product. The additive costs and technical support provided by a vendor should be determined before starting a trial usage program.

1.   General Environmental Science Corp.
      P.O. Box 22294
      Cleveland, OH 44122
      (216) 795-4733
2.   Ashland Chemical Co.
      P.O. Box 2219
      Columbus, OH 43216
      (614) 889-3333
3.   Polybac Corp.
      1251 South Cedar Crest Blvd.
      Allentown, PA 18103
      (215) 433-1711
4.   Sybron Corp.
      Biochemical Division
      Birmingham, NJ

III. BACTERIAL AUGMENTATION: EXPERIMENTAL BASIS

A laboratory study was needed to determine what basis, if any, existed for applying bacterial additives. The purpose of the study was also to determine the optimum aeration sequence to use in conjunction with the additive.

A. Background

The aerobic digestion process is similar to the activated sludge process. They differ in that activated sludge systems are used to degrade major organic and nitrogenous wastewaters whereas aerobic digestors are used to degrade organic sludges. In an aerobic digestor, the sludge is mixed and aerated and allowed to decompose. The sludge has very little food value and, once depleted, the microorganisms begin to consume their own protoplasm to obtain energy for cell maintenance. This results in volatile suspended solids (VSS) reduction, biological oxygen demand (BOD) reduction, an easily dewatered sludge, and a stable end product easily disposable. The main disadvantage to aerobic digestion is its high energy costs.

The MCB used consisted of seven strains of naturally occurring saprophytic bacteria marketed under the trade name Liquid Live MicroOrganisms (LLMO) and produced commercially by General Environmental Science Corp. The bacteria consist of both aerobic and anaerobic strains which have been isolated from soil and aquatic systems. The combination of bacteria of the LLMO was selected for their waste substrate-decomposing ability and their ability to cross-feed upon the by-products of each other's metabolism. The LLMO manufacturer claims that the addition of LLMO to a wastewater treatment system will increase BOD removal, reduce aeration needs, increase nitrification and denitrification, reduce solids, and eliminate obnoxious odors. The principle behind the effectiveness of LLMO in wastewater treatment plants is that the bacteria present in treatment systems are not the most effective microorganisms for organic decomposition. The source of bacteria for municipal wastewater treatment systems is the intestinal tract of man, an environment greatly different than that of a wastewater treatment system. LLMO bacteria are soil and aquatic bacteria whose environment is more closely aligned to the environment and needs of a wastewater treatment system.

Dramatic reduced aeration needs are claimed when LLMO is applied to aerobic digestors. This results because the LLMO manufacturer recommends changing the aerobic digestor into an aerobic/anaerobic digestor. The application of LLMO coincides with a gradual reduction in aeration over a 4- to 7-day period, culminating in about 3 to 6 hr/day aeration. This aerobic and anaerobic cycling provides the different strains of bacteria their most effective environment to attack the sludge. During the anaerobic stage, a number of processes occur:

- Fats, proteins, and carbohydrates are decomposed.
- Fatty acids are converted to carbon dioxide.
- Nitrate is converted to nitrogen gas and water.
- Hydrogen sulfide is converted to sulfate through anaerobic respiration and anaerobic photodecomposition.
- Sugars are fermented.

During the aerobic stage:

- BOD created from the anaerobic decomposition of fats, carbohydrates, and proteins is oxidized.
- Suspended solids are removed through the action of enzymes secreted by the bacteria.
- Hydrogen sulfide is oxidized to sulfate.
- Carbohydrates are decomposed to sugars.
- Sugars are oxidized to carbon dioxide.
- Ammonia is oxidized to nitrite.
- Nitrite is oxidized to nitrate.
- Proteins are broken down to peptides and amino acids.
- Peptides and amino acids are utilized for growth by most of the bacteria.

From these processes, it can be seen that the by-products of the LLMO are also substrates for microorganisms in LLMO.

B. Procedure

For the aerobic digestion study some 40 gal of sludge were obtained. The sludge was collected from the influent of the aerobic digestors of the Columbia Wastewater Treatment Plant (CWTP) and consisted of both primary and secondary sludge. The sludge was stored on ice en route to the TVA Unit Operations Lab (UOL) in Chattanooga, Tenn. and at 4°C for a 16-hr period at the UOL so that the test could be started on a timely schedule.

Six 5-gal reactors which model aerobic digestors were set up (see Figure 5). The differences simulated among the reactors were the amount of aeration applied and whether LLMO was added. All of the sludge samples were mixed and initial samples were taken. Approximately 5 gal of sludge were placed in each reactor. Reactors A and B modeled typical aerobic digesters with aeration being applied continuously for the duration of the study. Reactor A acted as the control for the system with no LLMO added. Reactor B had an initial dosage of 2 mg/ℓ of LLMO and a subsequent daily dosage of 1 mg/ℓ.

Reactors AA, BB, CC, and DD modeled the process design suggested for the application of LLMO for maximum energy savings and sludge reduction. The LLMO manufacturer

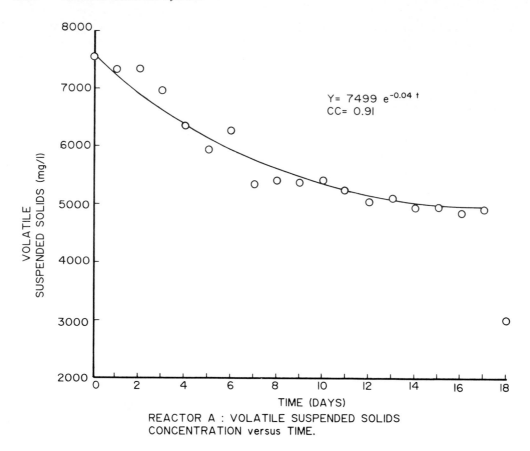

REACTOR A : VOLATILE SUSPENDED SOLIDS
CONCENTRATION versus TIME.

FIGURE 5.   Schematic of laboratory equipment.

suggested a gradual reduction in aeration over a 4-day period culminating in only 3 hr/day of aeration (1 hr per 8-hr shift). Three different aeration schedules were tried to determine the optimum aeration schedule for CWTP sludge. These schedules and a schematic of the study set-up are presented in Figure 4. LLMO was applied at the same concentration and schedule in Reactors BB, CC, and DD as previously discussed. No LLMO was added to Reactor AA, but it did follow the aeration on and off schedule suggested for the use of LLMO. Reactor AA acted as the control for Reactors BB, CC, and DD. Because of the different aerobic digestion process designs, different degrees of monitoring for controlling the systems were required. The straight aerobic digestion designs had pH and redox performed daily. DO was kept between 6 to 8 mg/ℓ and was measured intermittently. The aerobic/anaerobic system had sludge blanket level, pH, and redox performed at different times depending on the aeration schedule of the system. Hydrogen sulfide was performed on each aerobic/anaerobic reactor 1 hr before its particular anaerobic stage was to end. The H$_2$S sample was taken from the supernatant near the surface. Deionized water was added daily to compensate for the H$_2$S sample.

Design detention time for aerobic digestors is generally 15 to 20 days (Metcalfe and Eddy). The CWTP aerobic digestors have a design detention time of 16 days for decomposition of the sludge. It was estimated that the laboratory study would be finished in approximately 2 weeks. The same parameters as analyzed initially were analyzed at the completion of the study to determine the overall condition of each sludge and the efficiency of each reactor.

C. Results and Discussion

The volatile suspended solids (VSS) is the most important parameter in determining the

Table 2
SUSPENDED SOLIDS AND VOLATILE SUSPENDED SOLIDS
CONCENTRATIONS VS. TIME

Time (day)	A (mg/ℓ)		B (mg/ℓ)		AA (mg/ℓ)		BB (mg/ℓ)		CC (mg/ℓ)		DD (mg/ℓ)	
	SS	VSS	SS	VSS	SS	VSS	SS	VSS	SS	VSS	SS	VSS
0	9475	7540	9475	7540	9475	7540	9475	7540	9475	7540	9475	7540
1	8920	7320	8870	7240	8800	7200	9000	7390	9140	7500	9150	7460
2	9350	7330	9050	7150	8750	7020	8680	6970	8630	7010	8860	7170
3	8810	6970	8840	7060	8680	7110	8330	6820	8860	7290	8910	7330
4	8720	6350	8390	6090	9450	7090	9040	6680	8500	6650	8640	6890
5	8430	5920	8340	5820	8810	6430	8760	6280	8530	6200	9160	6720
6	8680	6270	8330	6050	8900	6660	8670	6330	8760	6510	8970	6650
7	7910	5350	7870	5420	8660	6150	8450	5890	8350	6120	8770	6290
9	7860	5400	7920	5430	8800	6300	8310	5800	8840	6290	8870	6350
9	7970	5370	7980	5270	8350	5830	8240	5600	8730	6090	9150	6400
10	7880	5420	7810	5330	8470	5930	8260	5660	8590	5950	9300	6440
11	7830	5240	7810	5230	8720	6250	8340	5760	8330	5860	8930	6270
12	7740	5070	7850	5090	8160	5720	7650	5130	8475	5790	8660	6090
13	7960	5110	7790	5070	8330	5680	8020	5370	7490	4870	9030	6290
14	7550	4950	7790	4990	8970	6140	8350	5520	8260	5670	8830	6140
15	7650	4970	7710	4920	8430	5680	8060	5380	8070	5540	9080	6270
16	7670	4890	7780	5040	8340	5590	7850	5160	8480	5840	8850	6040
17	7830	4940	7930	5130	8240	5460	8200	5430	7980	5440	8570	5880
18	7930	3030	7900	2950	8150	3560	7720	3060	7720	3180	8600	3890

efficiency and operation of an aerobic digestor. The VSS is the measure of the biologically active (viable) portion of the mixed liquor suspended solids (SS). Reduction of the VSS and SS is an indication that the sludge is being degraded. The VSS and SS vs. time data are presented in Table 2 and Figures 5 to 16. As can be seen from these figures, there is a large degree of variability in the data points. This is not unusual for either the VSS or SS test, especially in measurements of an organic sludge consisting of both primary and secondary clarifier sludge. It is extremely difficult to obtain a representative sample for these tests from a thick organic solution. Large particles and other debris may be picked up by the sampling pipette, impending sample integrity. All the reactors had extremely low values for VSS on the last day (18) of the study. This low VSS measurement did not have a corresponding trend in SS values. The VSS could not be degraded at this high rate in 1 day. The 18th-day value was considered erroneous and was not considered in the study (Table 3).

All the VSS vs. time data (Figures 5 to 10) seem to visually follow first-order removal kinetics. In a first-order reaction, the rate of decomposition is directly proportional to the amount of undegraded material (VSS). This may be expressed mathematically as:

$$-\frac{dx}{dt} = kx$$

where x = concentration of VSS in milligrams per liter, t = time of day, and k = rate coefficient for the reaction day^{-1}. The rate coefficient (k) is the slope of the line. The aerobic digestion Reactors A and B had the greatest VSS removal rate (Table 4) of all the reactors at 34 and 33%, respectively. The differences between the VSS results for Reactors A and B are insignificant considering the variability. The VSS rate coefficients for the aerobic/anaerobic reactors were 0.02 day^{-1}. The VSS removal rate for Reactors AA, BB, and CC was 28% and the removal rate for Reactor DD was 22%.

There is a great degree of variability in the SS vs. time data (Figures 11 to 16), and this

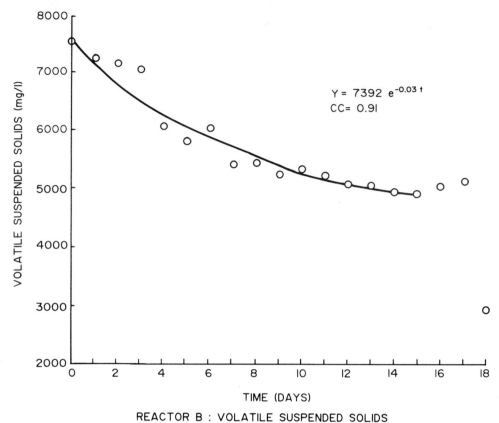

REACTOR B : VOLATILE SUSPENDED SOLIDS
CONCENTRATION versus TIME.

FIGURE 6. Experimental data.

is accountable in the SS test itself. First-order rate kinetics was applied to the data but was difficult to fit in some cases, and this is evident in the correlation coefficients.

The SS in the aerobic digestion Reactors A and B (LLMO) and the aerobic/anaerobic Reactor BB followed first-order removal kinetics best. The rate coefficients were 0.01, 0.02, and 0.01 day^{-1}, respectively. The rate coefficients of the other reactors were 0.01 day^{-1}, but the correlation coefficients were extremely poor.

The most dramatic results for comparing the operation of the reactors are presented in Table 3 which shows results of initial and final analyses, and Table 4, which shows percent reduction between initial and final analyses. The aerobic digestion Reactors A and B in almost all cases showed the greatest reduction. There was a slight difference between Reactors A and B with Reactor B showing a slightly greater overall percent reduction. This slight difference in most cases could be accounted for by analytical and sampling methods, not necessarily to the application of LLMO. Reactor DD had the lowest percent removals in almost all the parameters tested. These low removals were believed to be due to the state of upset the reactor seemed to be in until the final days of the study. Reactor DD had the lowest aeration time, 1-hr aeration per 8 hr. This aeration schedule seemed to be too low for the sludge to handle until the final days of the study.

A number of parameters had a great degree of difference between the aerobic digestion reactors (A and B) and the aerobic/anaerobic reactors (AA, BB, CC, and DD). These differences were mainly due to the different processes that occur between an aerobic and anaerobic process. Nitrite and nitrate were extremely high in the aerobic system but low in

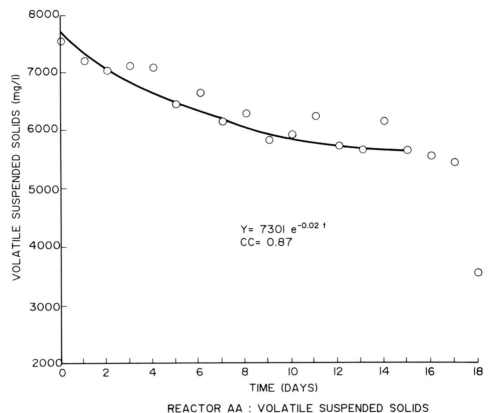

REACTOR AA : VOLATILE SUSPENDED SOLIDS
CONCENTRATION versus TIME.

FIGURE 7.   Experimental data.

the aerobic/anaerobic system. In both models, the ammonia was oxidized to NO_2 and NO_3, but only in the aerobic/anaerobic system was nitrate reduced to nitrogen gas and water. This reaction occurs in an anaerobic state.

The sulfide in the sludge was oxidized into sulfate in the aerobic system. This accounts for the 1900% increase in SO_4 in the aerobic system. The aerobic/anaerobic reactors did not show as great a sulfate conversion because of their anaerobic stage. In an anaerobic state (reducing), sulfates are reduced to sulfides and hydrogen sulfide (H_2S), and H_2S comes off as a gas. Of the aerobic/anaerobic reactors, the greatest H_2S production was shown by Reactor AA (no LLMO). All of the LLMO-treated aerobic/anaerobic reactors showed low H_2S production (see Tables 5 to 9), the lowest being Reactor BB which had the greatest aeration time. Reactor DD had the lowest aeration time and LLMO added, but still had less H_2S formation than Reactor AA. Suppression of H_2S formation during the anaerobic state by the LLMO-treated reactors may be the result of the ability of LLMO to convert H_2S to SO_4 under anaerobic conditions. Also for H_2S formation, a reducing environment is necessary, and LLMO-treated aerobic/anaerobic reactors had a tendency to remain for a longer period of time in the oxidizing state (a positive redox) following the aeration period. Settleable solids were measured using two methods: an Imhoff cone and a 1-ℓ graduated cylinder. These tests were used to evaluate and compare the settleability of the sludge. All of the reactors except DD had essentially the same settleability. Reactor BB had a slightly greater degree of settleability.

The pH of the aerobic/anaerobic reactors varied daily (see Tables 5 to 9) but stayed within the acceptable range of 6 to 8. The aerobic digestion reactors (A and B) required some slight

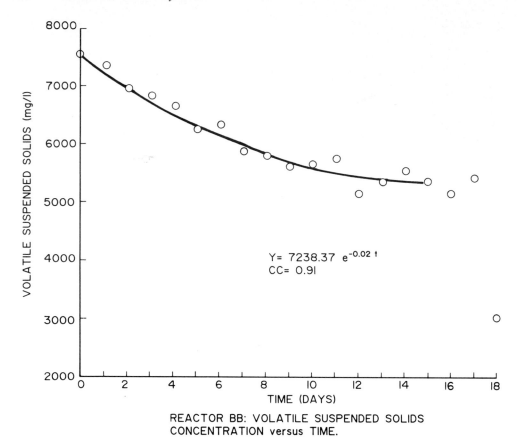

$$Y= 7238.37 \ e^{-0.02 \ t}$$
$$CC= 0.91$$

REACTOR BB: VOLATILE SUSPENDED SOLIDS
CONCENTRATION versus TIME.

FIGURE 8. Experimental data.

buffering beginning approximately at the ninth day of the study. Alkalinity is used when NH_4 is oxidized to nitrate in an aerobic digestor. No buffering is required by the actual plant aerobic digestors treating this sludge. It is not unusual to require additional buffering in small-scale laboratory setups.

The oxygen consumption rate of the reactors decreased significantly from the initial demand. This was expected because of the BOD and cell destruction which occurred over the 17-day test period. Reactors A, B, and BB had essentially the same reduction (89 to 91%) in oxygen consumption, which was slightly greater than the rest of the reactors.

The BOD and COD were reduced approximately the same across all of the reactors. There are some slight differences, but these may be explained by sampling and testing variables. The soluble BOD and soluble COD show good (>70 and >98% respectively) removals rates. No trends were evident.

Essentially the same percent removal of constituents was achieved between Reactors A, B, and BB. The aerobic/anaerobic Reactor BB had 47% less aeration than Reactors A and B. No discernible differences were evident by applying LLMO to straight aerobic digestion (Reactors A and B). Upset conditions were evident in Reactors AA, CC, and DD until late in the study. Reactor AA without LLMO could not handle the 68% reduced aeration applied to it. The supernatant was cloudy, and H_2S formation was prevalent. Reactor CC, at 69% reduced aeration, was upset slightly initially but cleared up after the 1st week of the study. Reactor DD had the highest reduced aeration at 80%. This reactor was in a state of upset until the late stages of the study. LLMO application could not compensate for the drastic aeration cut (3 hr/day) at the 4th day of the study. Reactor DD adjusted though toward the

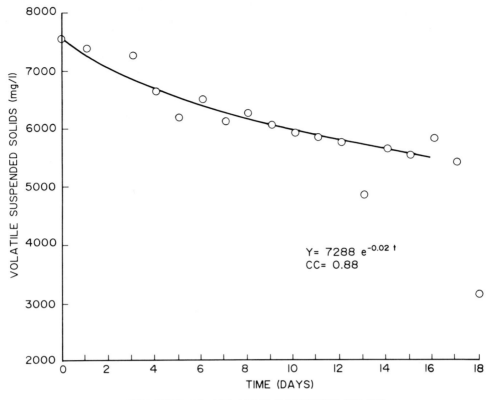

REACTOR CC: VOLATILE SUSPENDED SOLIDS
CONCENTRATION versus TIME.

FIGURE 9.   Experimental data.

end of the study. Upset conditions seem to prevail in reactors with drastic drops in aeration in the early stages of the study, but the LLMO-treated reactors adjusted in the later stages to the low aeration. This indicates that the LLMO-treated aerobic/anaerobic reactors can adjust to these low rates of aeration but that the aeration reductions need to be applied at a more gradual rate in the process. An extension of the 4-day gradual reduction in aeration to 8 to 10 days until applying the 3 hr/day of aeraton could result in the lowest aeration needs and approximately the same percent removals.

Since this study used batch digestors, the optimal aeration schedules for continuous-flow systems with decanting and sludge removal will probably be different. An engineering study should be performed for individual plants using the rate data of this study. A test program for the specific plant should be developed.

Linearization of the data for VSS and SS was performed as shown in Figures 17 to 23 with the results in Table 10. The following equations were used as digestor models:

$$VSS = VSS_0 \, e^{-k1 \, t}$$

$$SS \;\; = SS_0 \, e^{-k2 \, t}$$

Figure 24 is a photograph of two of the six lab-scale digestors.

D. Conclusions

1.   MCB applied to the aerobic digestion process without any process modifications had no significant effect.

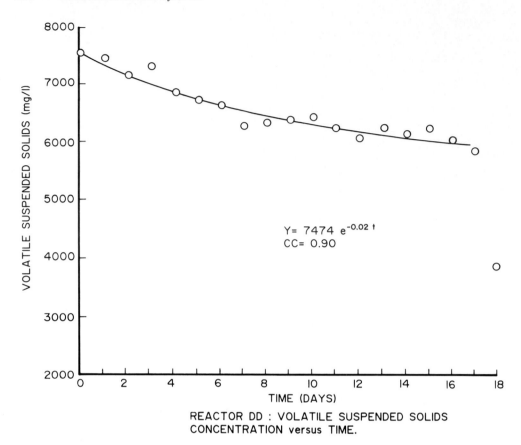

REACTOR DD : VOLATILE SUSPENDED SOLIDS
CONCENTRATION versus TIME.

FIGURE 10.   Experimental data.

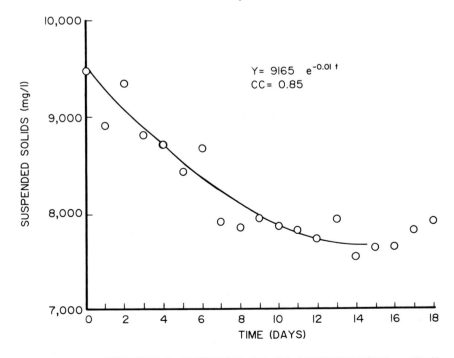

REACTOR A: SUSPENDED SOLIDS CONCENTRATION vs TIME.

FIGURE 11.   Experimental data.

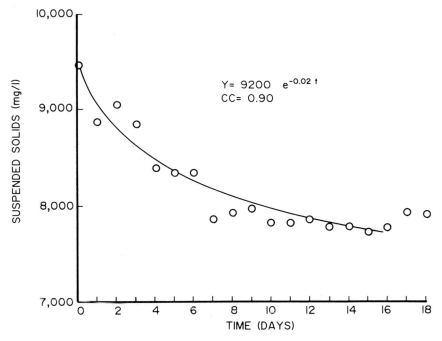

FIGURE 12.   Experimental data.

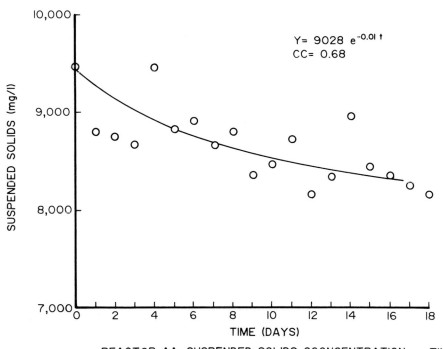

FIGURE 13.   Experimental data.

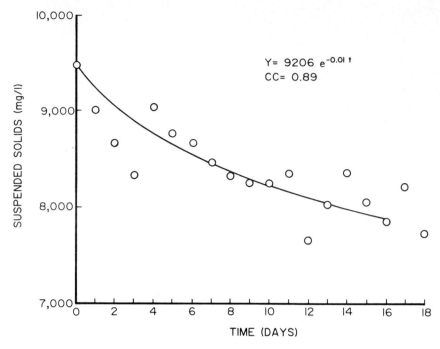

REACTOR BB: SUSPENDED SOLIDS CONCENTRATION vs TIME

FIGURE 14.   Experimental data.

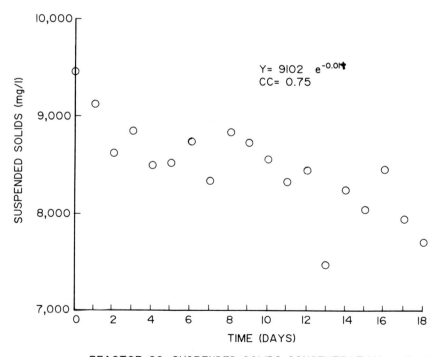

REACTOR CC: SUSPENDED SOLIDS CONCENTRATION vs TIME.

FIGURE 15.   Experimental data.

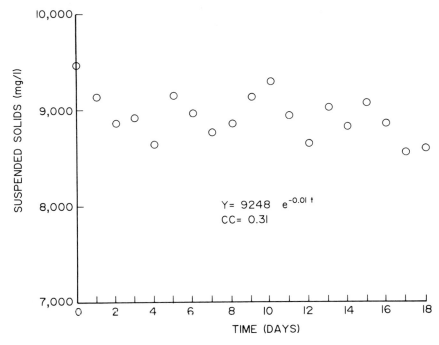

FIGURE 16.   Experimental data.

2. Approximately the same percent removal in constituents and settleability characteristics were achievable between the aerobic digesiton process and the MCB-treated aerobic-anaerobic digestion process with 47% reduced aeration.
3. The aerobic/anaerobic digestion process without MCB added had a large amount of H_2S formation, poor settleability, and a constant state of upset.
4. A more gradual and extended aeration reduction probably was needed in the aerobic/anaerobic process to allow the MCB microorganisms to adjust to the low supply of air.
5. Some optimal reduced aeration between 47 and 80% may be feasible for the MCB-treated aerobic/anaerobic process.

IV. DISSOLVED OXYGEN (DO) CONTROL

A. DO

As noted previously, the amount of air being used per digestor is less than design. This is due to the optimization of air usage in the plant. In the digestors, increasing the DO above 1.5 ppm by increasing the air flow has not worked. The diffusers used in the four digestors and in the aeration basins, preaeration basins, and sludge holding tanks are identical except for the number of diffusers per header. All diffusers used a $^5/_8$ in. orifice.

DO can be increased by increasing the oxygen transfer efficiency at constant COD. A low DO does not necessarily mean oxygen transfer is poor if the demand is high. Based on the observed patterns, supernatant, and VS data, however, it is believed that better digestor performance may be achieved if the DO can be increased to about 3 ppm. This is highly desirable to improve the overall performance of the bacterial additive being tested. Although the additive prevents hydrogen sulfide gas formation during the time when aeration is stopped, more rapid biological oxidation and nitrification is needed during the aerated portion of the

Table 3

RESULTS OF INITIAL AND FINAL ANALYSIS FOR ALL REACTORS

Parameter	Initial composite	A	B LLMO	AA	Final BB LLMO	CC LLMO	DD LLMO
COD, mg/ℓ	15,500	7700	7300	8200	7200	7900	8200
Soluble COD, mg/ℓ	245	73	52	59	54	53	56
BOD$_5$, mg/ℓ	5,200	2500	2400	2700	2700	3200	2400
Soluble BOD$_5$, mg/ℓ	190	1	<1	3	<1	4	2
NH$_4$ as N, mg/ℓ	39	0.18	0.19	0.45	0.40	0.27	0.31
TKN, mg/ℓ	67	94	96	81	75	92	83
NO$_2$ + NO$_3$ as N, mg/ℓ	0.08	160	130	2.1	3.3	1.9	1.1
SO$_4$, mg/ℓ	7	140	140	66	95	65	41
SS, mg/ℓ	9,475	7660 / 7930	7710 / 7900	8240 / 8150	7850 / 7720	7980 / 7720	8570 / 8600
VSS, mg/ℓ	7,540	4940[a] / 3030[b] / 4940[c]	5030[a] / 2950[b] / 4920[c]	5460[a] / 3560[b] / 5460[c]	5430[a] / 3060[b] / 5160[c]	5440[a] / 3180[b] / 5440[c]	5880[a] / 3890[b] / 5880[c]
Settleable solids, mg/ℓ (Imhoff cone)	925	775	785	760	740	750	800
Settled solids volume, mg/ℓ, graduated cylinder	700	440	435	450	440	450	570
Sludge volume index at 60 min	74	55	55	55	57	58	66
Oxygen consumption rate, mg/ℓ/min	1.575	0.1429	0.1524	0.3046	0.18	0.3396	0.3942
Specific oxygen consumption rate, mg/g/hr	12.53	1.74	1.81[c]	3.35	1.99	3.75	4.02
Suspended solids, removal coefficient, day^{-1}		0.01	0.02	0.01	0.01	0.01	0.01
Correlation coefficient		0.85	0.90	0.68	0.89	0.75	0.31
Volatile suspended solids removal coefficient, day^{-1}		0.04	0.03	0.02	0.02	0.02	0.02
Correlation coefficient		0.91	0.91	0.87	0.91	0.88	0.90

[a] VSS measurement on 17th day of study.
[b] VSS measurement on 18th day of study.
[c] Lowest VSS measurement of last 5 days, exluding last day.

Table 4
PERCENT REDUCTION BETWEEN INITIAL AND FINAL PARAMETERS

Parameter	Aerobic digestion		Aerobic/anaerobic digestion			
	A	B LLMO	AA	BB LLMO	CC LLMO	DD LLMO
COD, mg/ℓ	50	53	47	54	49	47
Soluble COD, mg/ℓ	70	79	76	78	79	77
BOD$_5$, mg/ℓ	52	54	48	48	38	54
Soluble BOD$_5$, mg/ℓ	99	99	98	99	98	99
NH$_4$ as N, mg/ℓ	99	99	99	99	99	99
TKN, mg/ℓ	−40[a]	−43[a]	−21[a]	−12[a]	−37[a]	−24[a]
NO$_2$ + NO$_3$ as N, mg/ℓ	−1999[a]	−1624[a]	−25[a]	−40[a]	−23[a]	−13[a]
SO$_4$, mg/ℓ	−1900	−1900	−800	−1300	−800	−500
SS, mg/ℓ	19	19	13	15	16	10
VSS, mg/ℓ	34	33	28	28	28	22
Settleable solids, ml/ℓ (Imhoff cone)	16	15	18	20	19	14
Settleable solids at 60 min	37	38	36	37	36	19
Oxygen consumption rate, mg/ℓ/min	91	90	81	89	78	75
Specific oxygen consumption rate, mg/g/hr	86	86	73	84	71	68

[a]   Negative signs indicate a percent increase from initial.

cycle. By generating more nitrate during aeration, the performance during anoxic conditions should improve. DO levels have been acceptable in the aeration basins.

According to manufacturers' literature, periodic inspections of the surface aeration pattern may warrant inspection of the diffusers. The surface pattern for the digestors will not be as good as that in the aeration basins due to a more dense mixed liquor in the digestors. While the diffusers of the aeration basin are submerged to about 13 ft, the digestor diffusers are only submerged to about 9 ft. Because the specific gravity is about 50% higher in the digestors, the system pressure remains balanced. Pressure drops across the diffusers in about 1 to 2 in. of water and across the entire diffuser assembly only about 2 in. Thus, a slight difference in the levels of the diffuser assembly ends can result in, say, the last section of diffusers being by-passed or more prone to clogging. Clogging is possible when the line pressure is less than the hydraulic head. Cycling of the aeration requires that the check valve and control valve are in good working order. Nonuniform air distribution may be caused by improper elevations of the diffuser assembly ends.

If the elevations of the various diffusers and diffuser assemblies are correct, the operating pressure can be increased by changing from $^5/_8$-in. orifices to, say, $^9/_{16}$-in. orifices in the diffusers. The system pressure will be increased by about $1^1/_2$ in. of water whenever the digestors are being aerated. The increased power consumption at the blowers may be compensated for if the oxygen transfer is enhanced by the expected finer bubbles in the digestors. This subject should be discussed with the diffuser manufacturer and the engineering consultants for the plant.

Because of the increased operating costs of aeration, it is suggested that the plant determine the economics of adding DO controls. A discussion of automatic controls for a 5-MGD plant follows.

B. DO Control System
An automatic control system to continuously monitor and maintain DO at a preset con-

Table 5
DAILY PARAMETERS MEASURED OF MODEL
AEROBIC DIGESTION A AND B

	A			B		
Day	pH	Redox (mV)	Dissolved oxygen (mg/ℓ)	pH	Redox (mV)	Dissolved oxygen (mg/ℓ)
1	7.3	+230		7.3	+270	
2	7.5	+220		7.2	+150	
3	7.6	+90		7.2	+60	
4	7.6	+260	7.0	7.1	+220	6.5
	7.6	+80		7.2	+200	
5	7.4	+280		6.9	+280	
	7.0	+260	7.4	6.5	+250	7.2
	7.1	+230		6.5	+150	
7	6.6	+290		5.7	+300	
	6.7	+200		6.5	+100	
9	6.0[a]	+320		7.2	+290	
	7.4					
10	7.0	+180		6.9	+240	
	6.8	+220		7.0	+150	
11	6.2[a]	+300	8.6	6.5[a]	+260	7.7
	7.0					
12	7.0	+280		7.0	+180	
	6.9	+280				
13	6.5	+310		6.3	+200	
	7.3	+330		7.3	+160	
15	7.0			7.1		
16	6.4	+320		6.5	+300	
	6.5	+330		6.7	+250	
	6.5	+300		6.6	+180	
17	6.2	+350		6.2	+310	
	6.3	+320				

[a]　Sodium bicarbonate was added for pH adjustment.

centration is relatively simple and cost-effective. Continuous DO sensors, or probes, monitor the DO concentration in the tank and send a signal, proportional to this concentration, to the controller (''black box''). The controller compares the present concentration level to the desired level and develops a corresponding control signal. This signal is then amplified and sent to the motorized control valve on the suction side of the aeration blower. The signal causes the valve to open or close in an amount proportional to the change in DO level in the tanks which is required. Variable speed drives may also be considered at higher costs.

The direct control of DO has been shown to result in substantial energy savings, on the order of 18 to 23% savings of total aeration energy. Automatic control of DO can be expected to be economically feasible for all activated sludge plants over 1 MGD capacity which have variable-flow influent streams.

DO control can be implemented in both diffused air and mechanically aerated activated sludge systems. In either case, DO probes are installed in the aeration tanks and continuously monitor the DO level. The probes pass a signal, the current being proportional to DO level, through the DO transmitters to a DO controller. In the case of a diffused air plant, the output of the controller is a voltage or air signal which controls the position of the inlet guide vanes or valve on the blower. This varies the rate of air flow through the blower. For mechanically aerated systems, the DO controller output is used to cycle or vary the speed of the aerators.

Table 6
DAILY PARAMETERS OF AEROBIC-ANAEROBIC REACTOR AA

Day	Hour	Sludge blanket[a] height (in.)	pH (s.u.)	Redox (mV)	H_2S (mg/ℓ)	Comments
1	10:00 a.m.	0[b]	7.2	0	0.55	Supernatant cloudy
		9				
	11:00 a.m.	0	7.3	−150	0.6	
		7.5	7.3	−240		
2	10:00 a.m.	0	7.3	−120		
		8.5	7.5	−180		
	11:00 a.m.	0	7.5	−240	0.55	Supernatant cloudy
		7	7.5	−250		
3	08:00 a.m.	c	7.8	+300		
	10:00 a.m.	0	7.3	−100	0.15	
	11:00 a.m.	8	7.4	−150	0.6	Supernatant cloudy
		0	7.3	−200		
		6	7.3	−250		
4	08:00 a.m.	c	7.7	+260		
	10:00 a.m.	0	7.4	−160		
	01:00 p.m.	7.375	7.4	−200	0.55	Supernatant cloudy
		0	7.4	−200		
		5.75	7.4	−230		
5	08:00 a.m.	c	7.7	+260		
	10:00 a.m.	0	7.4	−50		
	01:00 p.m.	7.75	7.3	−110	0.45	Supernatant cloudy
		0	7.4	−110		
		5.5	7.4	−140		
6	08:00 a.m.	c	7.6	+190		
		0	7.6	−60		
		7.375	7.6	−100		
		0	7.6	−60	0.25	
		5.375	7.6	−100		
7	08:00 a.m.	c	7.7	+250		
	10:00 a.m.	0	7.7	−30		
		7.1875	7.7	−50		
	11:00 a.m.	0	7.7	−50		
		6.125	7.7	−90		
	01:00 p.m.	0	7.7	+80	0.05	
		5.25	7.7	0		
9	08:00 a.m.	c	7.7	+260		
	10:00 a.m.	0[b]	7.9	+270		
		6.875	7.9	+160		
	11:00 a.m.	0	7.9	+180		
		5.875	7.9	+40	0—0.05	
	01:00 p.m.	0		+150		
		5				
10	08:00 a.m.	c	7.8	+120		
	10:00 a.m.	0	7.8	+180		
		6.375	7.8	+90		
	11:00 a.m.	0	7.8	+50	0	
		5.625	7.8	+50		
	01:00 p.m.	0	7.9	+75		
		5	7.9	+20		
11	08:00 a.m.	c	7.7	+275		
	10:00 a.m.	0	7.9	+270		
		6.5	7.9	+250		

Table 6 (continued)
DAILY PARAMETERS OF AEROBIC-
ANAEROBIC REACTOR AA

Time		Sludge blanket[a] height (in.)	pH (s.u.)	Redox (mV)	H$_2$S (mg/ℓ)	Comments
Day	Hour					
12	11:00 a.m.	0	7.5	+350	—	
		5.5	7.7	+250		
	01:00 p.m.	0	7.7	+250		
		4.875	7.7	+240		
13	11:00 a.m.	0	7.5	+350		Clear
		5.375	7.7	+290		
	01:00 p.m.	0	8.0	+260	0	
		5	8.0	+250		
16	08:00 a.m.	c	7.9	+370		
	10:00 a.m.	0	7.9	+300		
		6.375	7.9	+270		
	12:00 a.m.	0				
		5.125				
	01:00 p.m.	0	8.0	+300	0	Clear
		4.75	8.0	+250		

[a] Total water height equals 11.5 in.
[b] Zero sludge blanket height refers to measurements taken at the water surface during the anaerobic stage.
[c] Parameters at this time are taken right before aeration is stopped.

Table 7
DAILY PARAMETERS OF AEROBIC-
ANAEROBIC REACTOR BB

Time		Sludge blanket[a] height (in.)	pH	Redox (mV)	H$_2$S (mg/ℓ)	Comments
Day	Hour					
1	10:00 a.m.	8.5	7.2	+300	0	
	11:00 a.m.	0[b]	7.2	0	0.15	
		7	7.2	−50		
2	10:00 a.m.	0	7.0	+70		
		7.5	7.0	+30		
	11:00 a.m.	0	7.1	−90	0	
		6.5	7.1	−100		
3	08:00 a.m.	c	7.4	+260		
	10:00 a.m.	0	6.5	+50		
		7.5	6.6	−140		
	11:00 a.m.	0	7.0	−130	0	
		6.5	7.0	−190		
4	08:00 a.m.	c	7.1	+200		
	10:00 a.m.	0	6.5	+80		
		6.75	6.7	0		
	11:00 a.m.	0	7.0	−100	0.05	
		6.25	7.0	−200		
5	08:00 a.m.	c	7.0	+290		
	10:00 a.m.	0	7.0	+200		
		6.75	7.0	+110		
	11:00 a.m.	0	7.1	−30	0	
		6	7.1	−50		

Table 7 (continued)
DAILY PARAMETERS OF AEROBIC-
ANAEROBIC REACTOR BB

Time		Sludge blanket[a] height (in.)	pH	Redox (mV)	H$_2$S (mg/ℓ)	Comments
Day	Hour					
6	08:00 a.m.	c	7.6	+190		
	10:00 a.m.	0	7.6	−60		
		7.375	7.6	−100		
	01:00 p.m.	0	7.6	−60	0.25	
		5.375	7.6	−100		
7	0	c	7.0	+340		
	10:00 a.m.	0	7.3	+240		
	11:00 a.m.	6.625	7.3	+190	0	Extremely
		0	7.2	+220		clear
		5.75	7.2	+200		
9	08:00 a.m.	c	7.1	+300		
	09:00 a.m.	7.75				
	10:00 a.m.	0[b]	7.3	+300		
		5.875	7.3	+290		Clear
	11:00 a.m.	0	7.4	+200	0	
		5.375	7.4	+190		
10	08:00 a.m.	c	7.4	+160		
	10:00 a.m.	0	7.2	+250		
		5.875	7.3	+220		
	11:00 a.m.	0	7.4	+110	0	Clear
		5.375	7.4	−60		
11	08:00 a.m.	c	7.3	+220		Cloudy
	10:00 a.m.	0	7.4	+280		
		5.875	7.5			
12	11:00 a.m.	0	7.3	+220	0	
		5.125	7.3	+210		
13	11:00 a.m.	0	7.1	+340		
		5	7.1	+320		
	01:00 p.m.	0	7.4	+140		
		4.75	7.5	+150		
16	08:00 a.m.	c	7.5	+260		
	10:00 a.m.	0	7.6	+340		
		5.5	7.6	+340		
	12:00 a.m.	4.75				
17	08:00 a.m.	c	7.5	+350		
	10:00 a.m.	0	7.6	+330		
		5.875	7.7	+320		
	11:00 a.m.	0	7.2	+290		
		5	7.4	+290		

[a] Total water height equals 11.5 in.
[b] Zero sludge blanket height refers to measurements taken at the water
    surface during the anaerobic stage.
[c] Parameters at this time are taken just before aeration is stopped and
    no settling has occurred.

If multiple tanks are used, the individual speeds can be regulated according to the DO level in each particular tank. The mechanical aerators must be of the variable-speed type and sized to permit adequate mixing at reduced speeds while still maintaining solids in suspension.

Any of these schemes use simple, reliable analog control systems. There is no need to introduce digital control systems using computers or microprocessors. If, however, such

Table 8
DAILY PARAMETERS OF AEROBIC-ANAEROBIC REACTOR CC

Time		Sludge blanket[a] height (in.)	pH (s.u.)	Redox (mV)	H_2S (mg/ℓ)
Day	Hour				
1	10:00 a.m.	0[b]	7.1		0
		8.5	7.1	+230	
	11:00 a.m.	0	7.1	−30	0.15
		5.75	7.1	−110	
2	10:00 a.m.	c	7.1	−50	
		6.75	7.1	−230	
	01:00 p.m.	0	7.0	−100	0.2
		5.75	7.1	−110	
3	08:00 a.m.	c	7.2	+250	
	10:00 a.m.	0	6.6	−110	
		7	6.7	−150	
	01:00 p.m.	0	6.7	−150	0.3
		5.5	7.0	−180	
4	08:00 a.m.	c	7.1	+180	
	10:00 a.m.	0	6.9	−100	
		7	7.0	−110	
	01:00 p.m.	0	7.0	−80	0.2
		5.5	7.0	−200	
5	08:00 a.m.	c	6.9	+290	
	10:00 a.m.	0	7.2	−150	
		7.125	7.1	−140	
	11:00 a.m.	6.25			
	01:00 p.m.	0	7.0	−40	0.15
		5.375	7.1	−60	
6	08:00 a.m.	c	7.0	+170	
	10:00 a.m.	0	7.2	−50	
		7	7.1	−70	
	01:00 p.m.	0	7.1	−20	
		5.375	7.1	−50	0.15
7	08:00 a.m.	c	7.0	+300	
	10:00 a.m.	0	7.3	+40	
		—	7.3	−20	
	11:00 a.m.	0	7.2	+20	
		6	7.2	−20	
	01:00 p.m.	0	7.1	+100	0
		5.25	7.1	0	
9	08:00 a.m.		7.1	+280	
	09:00 a.m.	8.875			
	10:00 a.m.	0[b]	7.4	+200	
		6.5	7.4	+100	
	11:00 a.m.	0	7.5	+100	
		5.5	7.4	+60	
	02:00 p.m.	0		+150	
		5		+150	
10	08:00 a.m.	c	7.3	+130	
	10:00 a.m.	0	7.3	+60	
		6.125	7.4	+20	
	11:00 a.m.	0	7.4	0	
		5.625	7.4		
	01:00 p.m.	0	7.3	+220	0
		5	7.3	+110	
11	08:00 a.m.	c	7.2	+220	
	10:00 a.m.	0	7.4	+140	

Table 8 (continued)
DAILY PARAMETERS OF AEROBIC-ANAEROBIC REACTOR CC

Day	Hour	Sludge blanket[a] height (in.)	pH (s.u.)	Redox (mV)	H_2S (mg/ℓ)
		6.25	7.5	+70	
12	11:00 a.m.	0	7.1	+250	
		5.375	7.1	+200	
	01:00 p.m.	0	7.4	+240	
		4.875	7.4	+190	
13	11:00 a.m.	0	7.1	+300	
		5.5	7.2	+250	
	01:00 p.m.	0	7.5	+170	
		5	7.5	+100	
16	08:00 a.m.	c	7.3	+260	
		0	7.5	+300	
		6.125	7.5	+270	
	12:00 a.m.	5			
	01:00 p.m.	0	7.5	+140	
		4.75	7.5	+100	
17	08:00 a.m.	c	7.3	+320	
		0	7.5	+300	
		6.625	7.6	+290	
		0			
		5.5			
		0	7.5	+280	
		4.75	7.5	+250	

[a] Total water height equals 11.5 in.

[b] Zero sludge blanket height refers to measurements taken at the water surface during the anaerobic stage.

[c] Parameters at this time are taken right before aeration is stopped.

Table 9
DAILY PARAMETERS OF AEROBIC-ANAEROBIC REACTOR DD

Day	Hour	Sludge blanket[a] height (in.)	pH (s.u.)	Redox (mV)	H_2S (mg/ℓ)
1	10:00 a.m.	0[b]	7.1	−30	
		8.5			
	11:00 a.m.	0	7.0	−90	0.55
		7	7.1	−140	
2	10:00 a.m.	0	7.1	−150	
		6.75	7.1	−250	
		0	7.0	−100	
		5.5	7.0	−150	0.2
3	08:00 a.m.	c	7.1	+260	
	10:00 a.m.	0	6.8	−110	
		7	6.9	−150	
	11:00 a.m.	0	6.9	−160	
		6.25	6.9	−180	
	02:00 p.m.	0	6.9	−160	0.3
		5.5		−190	

Table 9 (continued)
DAILY PARAMETERS OF AEROBIC-ANAEROBIC REACTOR DD

Time		Sludge blanket[a]	pH	Redox	H₂S
Day	**Hour**	**height (in.)**	**(s.u.)**	**(mV)**	**(mg/ℓ)**
4	08:00 a.m.	c	7.0	+ 170	
	11:00 a.m.	0	7.0	− 175	
		6.75	6.9	− 200	
	01:00 p.m.	0	6.9	− 200	0.6
		5.5	6.9	− 220	
5	08:00 a.m.	c	7.1	+ 200	
		0	6.9	− 110	
		7	6.9	− 150	
		0	6.9	− 140	0.55
		5.875	6.9	− 200	
6	08:00 a.m.	c	6.9	+ 140	
	10:00 a.m.	0			
		7.75			
	11:00 a.m.	0	7.1	− 100	0.3
		6.625	6.9	− 160	
	02:00 p.m.	0	7.0	− 50	
		5.5	7.0	− 170	
7	08:00 a.m.	c	7.0	+ 190	
	10:00 a.m.	7.75			
	11:00 a.m.	0	7.1	− 50	
		6.5	7.1	− 90	
	01:00 p.m.	0	7.1	− 30	0
		5.375	7.1	− 60	
9	08:00 a.m.	c	7.1	+ 280	
	09:00 a.m.	9			
	10:00 a.m.	0[b]	7.3	+ 10	
		7	7.3	− 30	
	11:00 a.m.	0	7.4	+ 30	0—0.05
		6	7.4	0	
10	08:00 a.m.	c	7.2	+ 90	
	10:00 a.m.	0	7.4	− 40	
		6.875	7.4	− 50	
	11:00 a.m.	0	7.4	− 10	
		6	7.4	− 50	
	01:00 p.m.	0	7.4	+ 100	0
		5.125	7.3	+ 90	
11	08:00 a.m.	c	7.1	+ 190	
	10:00 a.m.	0	7.4	+ 70	
		6.75	7.4	− 40	
12	11:00 a.m.	0	7.4	+ 100	
		5.875	7.3	+ 30	
	01:00 p.m.	0	7.4	+ 160	
		5.0625	7.3	+ 140	
13	11:00 a.m.	0	7.1	+ 220	
		5.75	7.1	+ 150	
	01:00 p.m.	0	7.4	+ 240	
		5.125	7.4	+ 150	
16	08:00 a.m.	c	7.2	+ 360	
	10:00 a.m.	0	7.5	+ 190	
		6.625	7.5	+ 150	
	12:00 a.m.	5.5			
	01:00 p.m.	0	7.5	+ 250	
		5	7.5	+ 50	

Table 9 (continued)
DAILY PARAMETERS OF AEROBIC-
ANAEROBIC REACTOR DD

Time		Sludge blanket[a]	pH	Redox	H₂S
Day	Hour	height (in.)	(s.u.)	(mV)	(mg/ℓ)
17	08:00 a.m.	c	7.2	+260	
	10:00 a.m.	0	7.5	+240	
		7.25	7.5	+40	
	11:00 a.m.	6			
	01:00 p.m.	0	7.5	+220	0
		5	7.5	+30	

[a]   Total water height equals 11.5 in.
[b]   Zero sludge blanket height refers to measurements taken at the water surface during the anaerobic stage.
[c]   Parameters at this time are taken right before aeration is stopped.

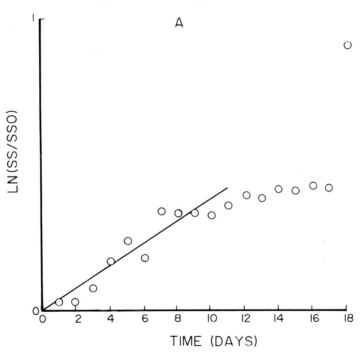

FIGURE 17.   Linearized rate data.

control equipment is already planned for the central operation of the plant, the DO control task can also be handled. Analog-to-digital and digital-to-analog converters would be required to interface the probes and control valves to the computer; the cost of this interfacing equipment would likely offset any savings realized by having the computer replace the DO controller.

One important criticism of DO control has been poor instrumentation reliability. In particular, fouling of the DO probes has been experienced frequently. Studies of users have reported that 20 to 25% of users found the probes unsatisfactory in 1977. Those users who

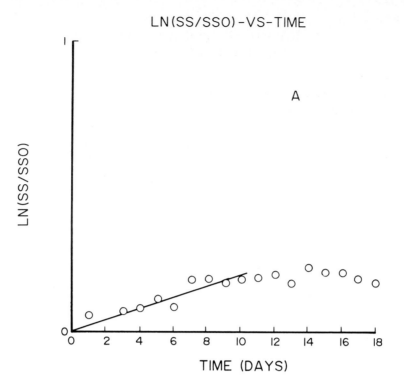

FIGURE 18.   Linearized rate data.

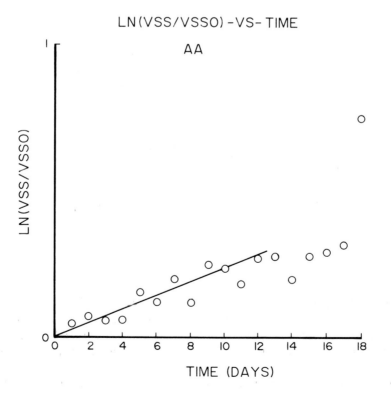

FIGURE 19.   Linearized rate data.

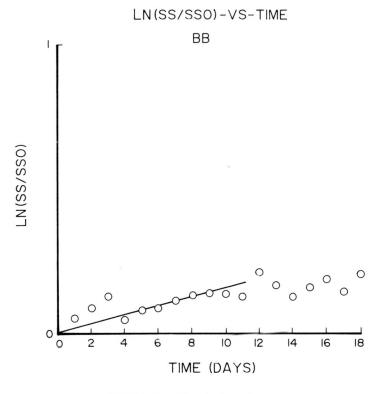

FIGURE 20. Linearized rate data.

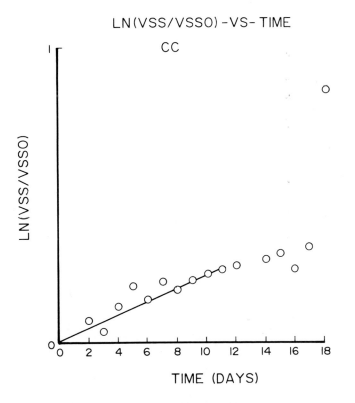

FIGURE 21. Linearized rate data.

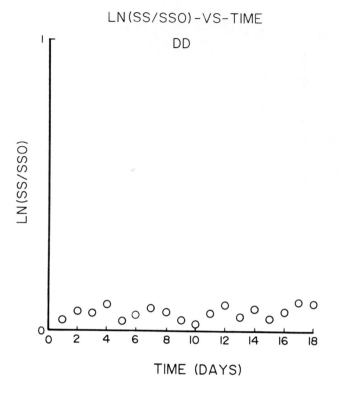

FIGURE 22.   Linearized rate data.

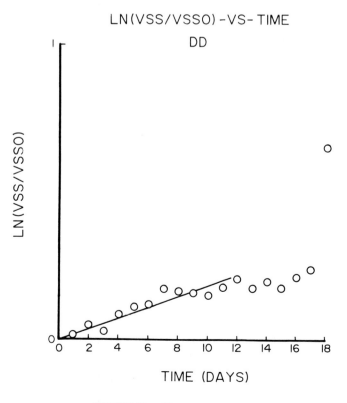

FIGURE 23.   Linearized rate data.

Table 10
SUMMARY OF DECAY-RATE
CONSTANTS

Reactor		K1 (day)	K2 (day)	% Air reduction actual	hours
A	control	0.02	0.035	0	432
B		0.02	0.038	0	432
AA	control	0.005	0.022	68	138
BB		0.015	0.03	47	228
CC		0.01	0.025	69	132
DD		0.001	0.02	80	87

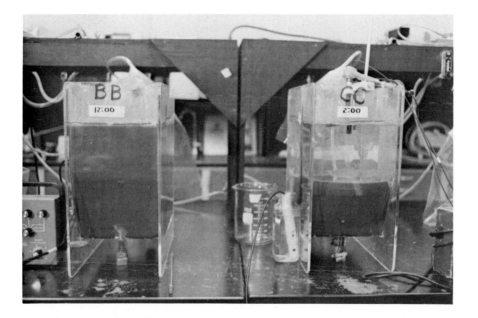

FIGURE 24.   Laboratory digestors.

experienced success with the probes suggested daily-to-weekly probe inspections and weekly calibrations. Membrane fouling was the chief maintenance problem with failures occurring from 1 to 9 months apart. Recharging of electrolyte is required every 3 to 9 months, depending on the model probe used.

Manufacturers claim that the probes being marketed today are more reliable. In addition, they are willing to support municipalities in training and regular maintenance programs. Before investing in a DO control system, the municipality should evaluate several different manufacturers and determine a maintenance schedule that operators agree to perform.

The capital cost of DO control systems is high. Retrofitting an existing installation is likely to cost about $25,000 for a 5 MGD plant, with energy savings of about $8000 per year. Because of the additional operations and maintenance expenses incurred, the payback period is on the order of 5 years (see Table 11).

Table 11
DISSOLVED OXYGEN CONTROL ECONOMICS

	5-MGD Plant
Capital Costs	
DO probes	$7,500
DO probe receptacles	2,000
DO transmitter	6,000
DO recorder	6,000
DO controller	1,000
Motorized control valve	500
Miscellaneous	2,000
Total	$25,000
Operations and maintenance costs (annual)	
DO probes (parts and labor)	$2,000
DO controls and instr. (part and labor)	1,000
Operations checks and calibration	1,000
Manual control labor savings	(1,000)
Energy savings (annual)	$8,000
Total annual savings	$5,000
Simple payback period	5 years

V. RECOMMENDATIONS FOR AN ADDITIONAL PLANT AERATION SYSTEM

A. Description of Aeration System

The present volume being aerated is as follows:

Seven aeration basins, total volume = (7)(253,703 gal) = 1,775,921 gal = 237,423 ft^3
First aerobic digestor volume = 253,703 gal = 33,918 ft^3
Second aerobic digestor volume = 211,978 gal = 28,339 ft^3
Total aerated volume = 300,000 ft^3

The plant has three 200-hp and two 75-hp turboblowers. Data from 8/1/82 to 8/10/82 show that the plant used one 75-hp plus one 200-hp turboblower to provide aeration. The amount of power consumed by the blowers normally fluctuates when outside temperature variations require more or less air to keep DO levels above acceptable minimums. Each turboblower has a minimum operating point to avoid surge and a maximum operating point to prevent motor overload. These conditions were estimated and summarized as follows:

75-hp Turboblower:	480 V; 3 phase; 88% power factor
Surge point:	46 Å; 35 kW; 600 scfm
Representative op. point:	68 Å; 51 kW; 1200 scfm
Max. capacity:	80 Å; 60 kW; 1500 scfm
200-hp Turboblower:	480 V; 3 phase; 88% power factor
Surge point:	125 Å; 93 kW; 1500 scfm
Representative op. point:	185 Å; 138 kW; 3200 scfm
Max. Capacity:	214 Å; 160 kW; 4000 scfm

For the representative operating point the aeration system provides 4400 scfm, which provides about 14.7 scfm per 1000 ft^3 of aerated volume. The actual amount needed depends on the dissolved oxygen concentrations, which are kept between 1 and 2 ppm.

B. Recommendations for Energy Savings

TVA laboratory- and full-scale tests have shown that it is possible, when using commercially available liquid culture bacterial additives (MCB), to reduce the amount of digestor aeration. The amount of aeration, the amount of additive, and the frequency of dosage depends on the digestor design and sludge properties of the individual plant. As a preliminary guide, the amount of MCB per dose for digestors is 1 to 2 ppm. The frequency of dosage depends on the flow scheme. For a batch operation, the frequency may be less than that for a continuous-flow system. Daily dosage is recommended for a continuous-flow system. The plant must make laboratory measurements of digestor performance, including nitrate levels, to determine if the additive is effective. An important function of the additive should be to utilize nitrate during anaerobic conditions when aeration is interrupted. This probably prevents hydrogen sulfide formation from sulfates entering the disgestors. According to lab studies, the MCB method will not work unless the mixed liquor is subjected to periods or zones of alternating aerobic-anaerobic conditions. This can be achieved in a continuous-flow system by providing zones where air diffusers are shut off and dispersion is minimized by baffles or partitions. Because of mixing considerations, this may not be practical. Attempting to use basins in series where air is completely cut off in individual basins will probably result in flow and mixing problems. If mechanical mixers are used in combination with separate aeration equipment, this may be practical. For plants that have a continuous-flow system that uses diffusers for aeration and mixing it is suggested that all the digestor basins have their aeration cycled on a sequential schedule, such as the following example for a 2-blower system.

Beginning each shift, the operator would find both digestor basins with no air flow. He would open the air valve of the first digestor to the operating point (about 500 scfm). After about 2 hr of aeration, he would close the air valve on the first digestor and then open the air valve on the second digestor to its operating point (400 scfm). He would close this valve after about 2 hr. Throttling of the 200-hp blower intake may help if surge conditions develop on the 75-hp blower when the digestors are not aerated. Throttling of blowers is preferred over frequent starting and stopping of the blower motors. Motor efficiency drops rapidly as the load drops below 50%. Power factor is even more sharply reduced.

The plant may find that the 75-hp blower can be shut off and only a single 200-hp blower used to provide aeration for the digestors and the aeration basins. A conservative estimate is made based on the aeration schedule of Figure 25 using both a 75- and a 200-hp blower. The laboratory staff would have the responsibility for computing the amount of additive needed and adding it to the first digestor during aeration on the first shift. For example, if the volume in the first digestor was 250,000 gal, then a 2-ppm concentration would require that 0.5 gal be added. Using a typical cost per gallon of $14, the annual cost of 0.5 gal/day would be $2,555.

From Figure 25, the demand will be reduced from 189 to 168 kW for a saving of 21 kW. The energy consumption will be reduced from 1,655,640 kWhr to 1,370,940 kWhr for a saving of 284,700 kWhr on an annual basis. The dollar savings is computed as

$$(21 \text{ kW})(\$6.84 \text{ per kW per month})(12 \text{ months per year})$$

$$+ \quad (284,700 \text{ kWhr})(\$0.02614 \text{ per kWhr})$$

$$= \quad \$9,166$$

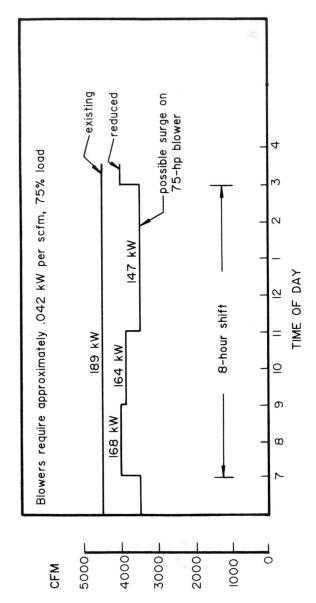

FIGURE 25. Plant aeration schedule.

The existing conditions of 189 kW are taken as an average for illustrative purposes only. The actual demand will be different for each month as reflected in previous billing data.

To modify the previous idea, the 75-hp blower would be shut off. This places demand for air totally on one 200-hp blower and relieves the operator of attending to surge conditions. The maximum output of the 200-hp blower is about 4000 scfm ($\sim$ 160 kW). When the disgestors would not be aerated, the air required is 3500 cfm (147 kW). When air is supplied to the digestors, the blower will be fully loaded. If the load is excessive, some reduction in air to the aeration basins during digestor aeration can probably be tolerated. The kilowatt and kilowatt hour savings are taken to be the same, although the efficiencies of the larger and smaller motors will result in differences.

To further reduce the blower power, air flows less than that required to maintain 14.7 cfm/1000 ft^3 of liquid can be attempted. If DO levels stay above 1 ppm, then the air flow is probably sufficient. Clogging of diffusers is a potential problem that is present with or without cyclic aeration. Cycling diffusers off will aggravate a clogging or plugging condition. A trial period is needed to determine if cyclic operation can be implemented.

ACKNOWLEDGMENT

The author wishes to acknowledge the support of the City of Columbia, Tenn. and the Divisions of Energy Conservation and Rates and Natural Resource Services of the Tennessee Valley Authority. Mr. Jeffery J. Longaker of the TVA Water Quality Branch conducted the experimental evaluations.

REFERENCES

1. **Reid, Crowther and Partners, Limited,** Designing an energy efficient wastewater treatment plant. I—III, *Water Sewage Works,* November 1979 to January 1980.
2. **Stahl, J. F. et al.,** Energy conservation in the design and operation of wastewater treatment facilities, *Manual of Practice,* No. FD-2, Water Pollution Control Federation, Washington D.C., 1981.
3. **Keith, D. A., et al.,** Energy efficiency in water and wastewater treatment plants, Georgia Institute of Technology EES, Project A-2836, Office of Energy Resources, Atlanta, Ga., February, 1981.
4. **Rucker, D.,** Bacteria cut power bills for sewage sludge, IMPACT, 5(3), 6, 1982.
5. **Reed, M. W.,** Energy conservation in aerobic digesters, TWWA Middle Section Meeting, Columbia State Community College, 1984.
6. **Horsfall, F. L., III,** Biochemical augmentation of wastewater treatment, *Deeds and Data,* Water Pollution Control Federation, Washington, D.C., 1977.
7. Communication with General Environmental Science Corp., P.O. Box 22294, Beachwood, Ohio, 44122.
8. USDE, Energy optimization utilizing bacterial augmentation, U.S. Department of Energy, Washington, D.C., March, 1983.
9. **Horsfall, F. L., III,** Bacterial augmentation of wastewater treatment, *J. N. Engl. Water Pollut. Control Fed.,* 13 (No. 2), 158, 1979.
10. **Anon.,** Bugs reduce electrical consumption, *Electr. Constr. Maint. J.,* 38, 1981.
11. **Morgan, J. R.,** Aerated lagoon well suited for village of Tuscarawas Ohio, *Ohio WPCC Buckeye Bull.,* 52 (No. 4), 1979.

Chapter 5

LOW-COST ANAEROBIC DIGESTION OF PIG MANURE AT PSYCHROPHILIC TEMPERATURE

J. Rieradevall, A. Ruera, L. Postils, and M. Vicente

TABLE OF CONTENTS

I. AIMS AND OBJECTIVES

The object of this preliminary study is to evaluate the possibilities of using low-cost anaerobic systems, which work at low temperatures, in intensive pig farms in Spain.

First of all, in order to verify whether or not the anaerobic digestion was possible in our lattitudes without the help of outside heating (low temperature or psychrophilic digestion), a study of the temperature behavior of a standard pig manure pit was carried out over the period of 1 year.

According to the positive results of the study (temperatures were always above 10°C and daily oscillations were ± 1°C, which could make a psychrophilic fermentation possible), the design of the modifications to be made to the pit for turning it into a low-cost digester has been developed along with an estimate of the cost of the project as a whole, taking care that the only autoctonous material and easy-to-manage equipment should be used. Finally, an estimate of the probable economic results of the project was made, comparing revenues (considering as such the savings derived from substituting conventional energy sources used in the farm for heating purposes, by the biogas produced) and costs (installation repayment and maintenance).

II. INTRODUCTION

Low-cost conventional continuous digestions developed in India, China, and Taiwan (see Figures 1 to 3) have the following general characteristics: local methods are used to construct the digesters, the volume of the digester is low, the loading system is semicontinuous, and the whole does not have insulation or any heating.[1-4]

In Europe, the study and development of low-cost and low-temperature (psychrophilic) digesters has been focused on the reconversion of existing manure pits into digesters (see Figure 4) with the finality of solving the economic problems deriving from the high cost of the installations and low level of net energy output of conventional digesters.[5,6]

If we study the list of approximately 25 digesters now in operation (1983) in Spanish farms, the following points arise:

1. The technology applied in a great deal of them is foreign.
2. Most are expensive and sophisticated installations with a high cost per cubic meter of digester, particularly in the medium and small digesters (see Figure 5).
3. The range of working temperatures is 30 to 40° C (mesophilic).
4. No details are available concerning the energetic balance and the economic results of the operations.

All these factors act adversely upon the process of introducing this new technology in medium and small farms, which represent the majority of those found in Spain. Faced with these problems, it appeared to be necessary to carry out a study of developing a new technology which would allow the adaptation of the existing anaerobic digestion know-how to the particular characteristics of our pig farms — an intermediate level technology with a more favorable energetic balance. We have tried to develop our research along these lines, with the following schema:

1. Carrying out a temperature study in a pig manure pit in order to determine the possibility of applying systems which work at low (psychrophilic) temperatures (between 10 and 25°C).
2. Drawing up a preliminary design of low-cost digester adapted to the medium and small Spanish pig farms.
3. Making an economic study.

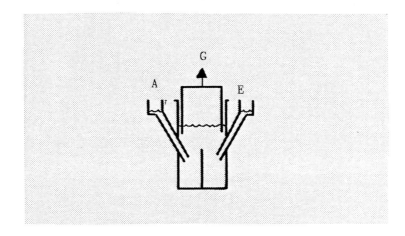

FIGURE 1. Continuous, conventional Indian type A, affluent; E, effluent; G, biogas.

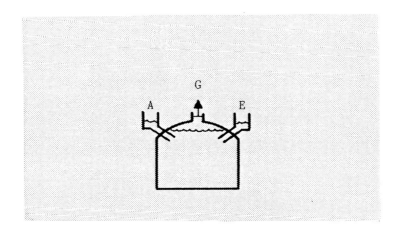

FIGURE 2. Continuous conventional Chinese type. A, affluent; E, effluent; G, biogas.

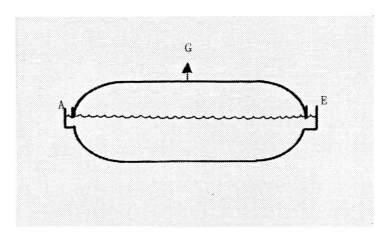

FIGURE 3. Continuous conventional Taiwan type. A, affluent; E, effluent; G, biogas.

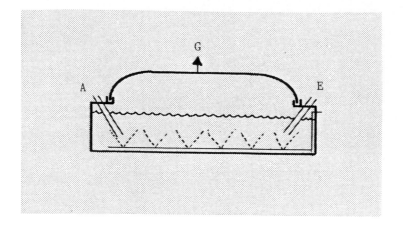

FIGURE 4.   Manure pit transformed into a low-cost digester. A, affluent; E, effluent; G, biogas.

III. TEMPERATURE STUDY IN THE PIT

A. Background

Anaerobic digestion occurs in nature within a wide range of temperatures, from 4 to 90°C: at 4°C in the glaciers, from 5 to 15°C in the marine sediments and lakes, at 30°C in the anaerobic digestion plants, at 39°C in the ruminant's paunch, and from 60 to 90°C in the geysers.[7,8] The temperatures at which the anaerobic digesters can work are psychrophilic processes: from 10 to 25°C; mesophilic processes: from 30 to 45°C; thermophilic processes: from 50 to 65°C. Of these, the mesophilic process is the one most commonly used in the installations which deal with pig manure.

The interest in carrying out fermentation at psychrophilic temperatures rather than at mesophilic ones is due to trying to make the energetic balance of the digesters optimum, since 90% of the energy they consume is used to raise the initial temperature of the residues to be processed, and only 10% is used to offset the losses in temperature of the digester itself. It is thus understandable that the lower the initial temperature, the lower the total energy consumed. At the same time, it has been observed that upon working at lower temperatures, the fermentive process is less sensitive to temperature fluctuations and loading fluctuations in the digester, control costs diminish, antibiotic treatment of the animals is permitted if necessary (which could otherwise stop the fermentation process), and free ammonium concentration is lower, since the temperature is lower and inhibition diminishes. In spite of these positive aspects, there are also some negative ones, such as: the gas production rate diminished with the temperature[10] and a bigger digester is necessary as the hydraulic retention time (HRT) to produce the same level of degradation as in the mesophilic process is more than doubled.

B. Methods and Results

For 1 year, the temperature of the contents of a manure pit were registered at different depths, and a record of outside temperatures was also kept. The pit used in this study is underground and covered, measuring 7 m × 7 m × 2 m in height, with 20-cm concrete walls. It is situated next to a pig operation at the Farming School "Torre Marimon", at Caldes de Montbui, Barcelona, Spain (with a closed cycle of 50 sows). Manure production is 2000 ℓ/day (see Figure 6).

Three temperature sensors (maximum 50°C and minimum −30°C) were used, placing them on a mobile support situated at the center of the pit at three different distances from

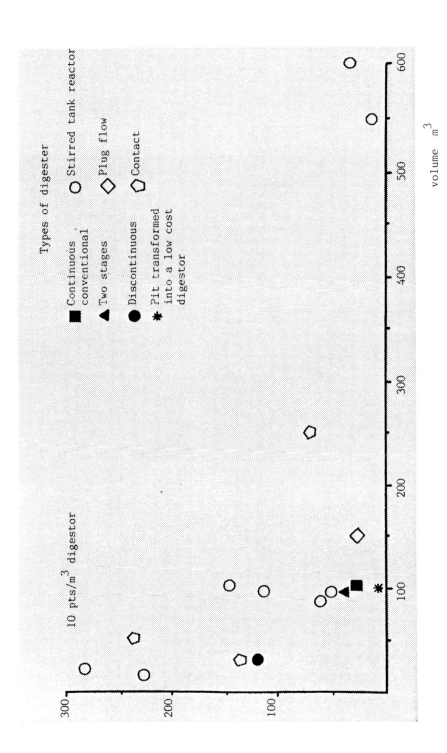

FIGURE 5.   Cost per cubic meter of digester in present Spanish anaerobic digestion installations according to the total volume of the digester. 1$ = 170 ptas.

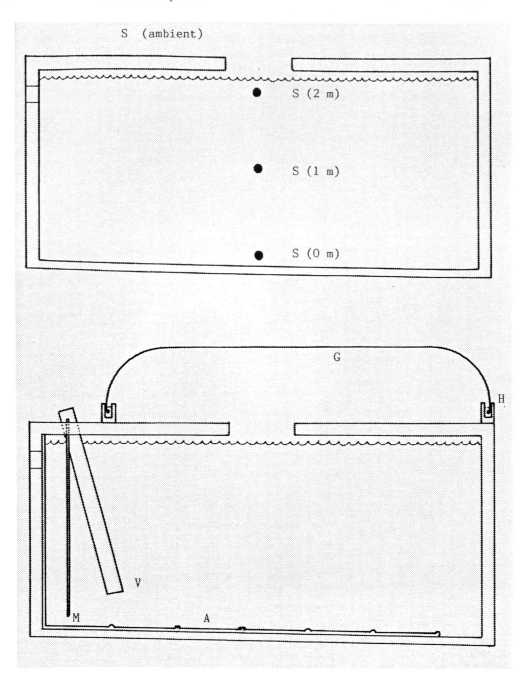

FIGURE 6.   Above — storage pit of pig manure. Below — storage pit transformed into a low-cost digester. S = temperature sensor, A = pneumatic shaker, G = gasometer membrane, H = hydraulic security valve, V = pit emptying pipe, M = sampler and sensors.

the bottom of the same: from 0, 1, and 2 m. The temperatures inside the pit at the three depths, as well as the outside temperatures, were registered periodically from September 1983 to August 1984. The results can be seen in Tables 1 and 2. Table 3 shows the average values, standard deviation, and the variability rate of the aforementioned results.

C. Discussion

From the data obtained, considering both the pit alone and also in relation to the sur-

Table 1

TEMPERATURE ANALYSIS IN THE PIT FROM SEPTEMBER 1983 TO AUGUST 1984

Height from bottom	0 min				1 min				2 min			
	$\overline{T}_{max}$	$\overline{T}_{min}$	Θ	n	$\overline{T}_{max}$	$\overline{T}_{min}$	Θ	n	$\overline{T}_{max}$	$\overline{T}_{min}$	Θ	n
September	25.70	22.90	2.80	14	24.60	23.10	1.60	14	27.80	23	5.20	13
October	24.90	22.00	2.80	22	24.60	22.00	2.50	22	27.00	23	4.30	7
November	19.30	17.10	2.40	30	17.00	14.30	2.30	27	V	V	V	—
December	17.10	15.40	1.70	25	16.50	14.90	1.60	25	V	V	V	—
January	14.00	12.70	1.20	18	13.60	12.70	0.72	18	V	V	V	—
February	12.10	10.50	1.70	22	11.80	9.60	2.20	12	V	V	V	—
March	14.20	10.50	4.00	4	13.25	9.50	3.80	4	V	V	V	—
April	12.70	11.80	1.70	10	R	R	R	—	13.60	12	1.90	12
May	16.80	14.60	2.40	10	18.00	14.70	3.40	16	18.70	14.80	4	16
June	22.70	15.20	7.50	10	21.70	15.40	6.10	14	21.40	15.90	5.50	8
July	21.86	21.00	1.00	18	22.16	21.16	1.20	18	23.86	21.80	2.10	18
August	23.40	22.10	1.10	4	22.80	21.80	1.34	4	V	V	V	—

Note: V = The level does not reach 2 m. R = Sensor breakdown. n = Number of samples taken. $\overline{T}_{max}$ = Average maximum temperature. $\overline{T}_{min}$ = Average minimum temperature. Θ = Daily fluctuation.

Table 2

MAXIMUM, MINIMUM, AND AVERAGE TEMPERATURES IN THE PIT AND ON THE OUTSIDE FROM SEPTEMBER 1983 TO AUGUST 1984

Month	Pit				Atmosphere			
	$\overline{T}_{max}$	$\overline{T}_{min}$	$\overline{T}_{media}$	Θ	$\overline{T}_{max}$	$\overline{T}_{min}$	$\overline{T}_{media}$	Θ
September	26.03	23.00	24.50	3.20	28.50	14.60	21.55	13.47
October	25.50	22.30	23.90	3.20	23.00	9.50	16.25	13.47
November	18.15	15.65	16.90	2.35	17.50	8.30	12.90	8.73
December	16.80	15.15	15.98	1.65	13.90	0.80	7.35	13.11
January	13.80	12.70	13.25	0.96	12.00	0.30	6.70	12.66
February	11.95	10.05	11.00	1.95	13.10	0.40	6.00	12.71
March	13.73	10.10	11.90	3.90	14.50	0.96	8.00	13.48
April	13.15	11.90	12.53	1.80	18.50	5.10	11.70	13.47
May	17.83	14.70	16.28	3.27	19.40	7.50	13.30	12.71
June	21.93	15.50	18.72	6.37	25.60	13.60	18.40	13.70
July	22.61	21.32	21.97	1.43	30.40	16.50	13.50	13.87
August	23.10	21.95	22.53	1.20	28.20	15.56	21.89	13.05

Note: $\overline{T}_{max}$ = Average maximum temperature. $\overline{T}_{min}$ = Average minimum temperature. T_{media} = Average temperature. Θ = Average daily temperature.

rounding atmospherical temperature, we come to some conclusions, which are discussed next.

1. Pit

The average monthly temperatures are similar at each of the three levels throughout the year (the greatest difference is 3°C). Minimum temperatures have the same pattern (Table 1).

Table 3

STATISTICAL TEMPERATURE STUDY

Height	Max.F 0 m	Min.F 0 m	Max.F 1 m	Min.F 1 m	Max.F 2 m	Min.F 2 m	Max.F media	Min.F media	Med.F	Max. amb.	Min. amb.	Media amb.
$\bar{x}$	18.73	16.31	18.72	16.28	22.06	18.41	18.71	16.19	17.37	20.38	7.76	13.96
σ	4.68	4.50	4.44	4.72	4.89	4.35	4.78	4.58	4.62	6.28	6.00	6.04
γ	24.99	27.59	23.73	28.99	22.17	23.63	25.57	28.34	26.60	30.81	77.41	43.27

Note: $\bar{x}$ = Average; σ = standard deviation; γ = variability rate; F = Pit.

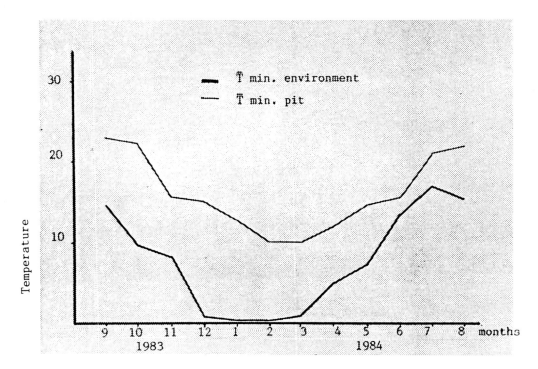

FIGURE 7.   Average minimum monthly temperatures.

We consider that the average values at 2 m are somewhat distorted, due to lack of data, because the manure has been used in a normal way and has perperiodically been taken to the fields, so that the level in the pit did not always reach the 2-m level.

2. *Relationship of Pit/Environment Temperatures*

The average minimum temperature at the pit did not fall below 10°C during the 12 months of the study, whereas outside it was lower than 1°C from December to March (see Figure 7). The average minimum outside temperature was 7.7°C whereas in the pit it was 16.2°C; therefore, with a very significant difference at 8.5°C. The variability rate of the minimum outside temperature was 2.73 times greater than in the pit itself (Table 3). On the other hand, the maximum temperatures inside and out show a certain parallelism, since the variability rate of the outside maximum temperature is only 1.2 times greater than in the pit (see Figure 8).

With regards to the average temperatures, we can see that in the pit there are two different periods: the first one runs from December to March (winter), with temperatures between 10 and 15°C; the second one has temperatures over 20°C from July to September (summer) (see Figure 9). Outside daily average temperature oscillation is around 13°C, significantly higher than in the pit itself which is only of 2°C (see Figure 10).

IV. LOW-COST DIGESTER DESIGN

A. Background

The use of mesophilic anaerobic digesters in small and medium pig farms in Spain has been hampered by the high installation cost and the low net energy output. To lower the consumption of energy in the process, two possible methods exist:

1.   Use of a heat exchanger between the manure intake and the effluent.
2.   Adoption of low-temperature systems.

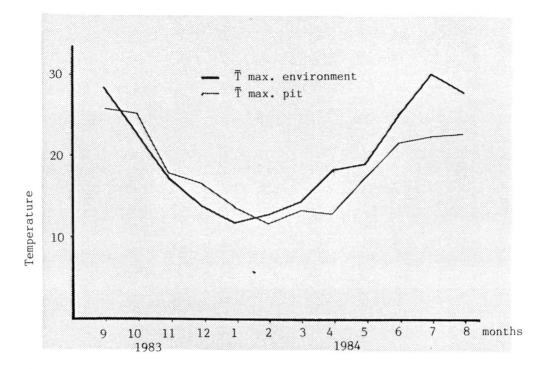

FIGURE 8.   Average maximum monthly temperatures.

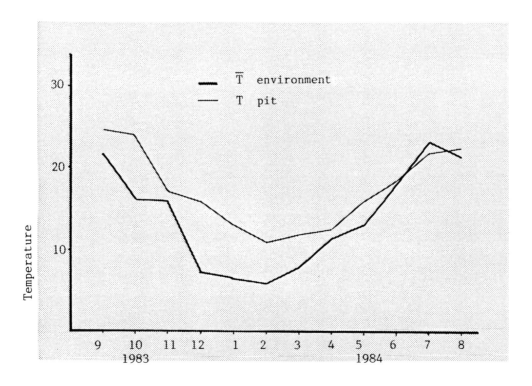

FIGURE 9.   Average monthly temperatures.

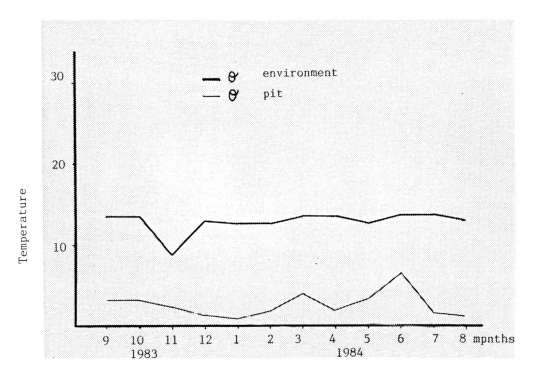

FIGURE 10.   Monthly temperature fluctuation.

Method 1 can have functional defficiencies due to the heterogenous composition of the manure (particularly the clogging and the poor temperature transmission) and Method 2 demands a larger digester volume.

The temperature follow-up aforementioned has allowed us to see that the psychrophilic fermentation would be possible, technically, throughout the year in our latitudes. In another study carried out on farms in Catalonia (in the northeast of Spain), the following conclusions have been drawn: 76% of the pig manure storing pits are covered, the most usual building material is concrete. The relationship of surface/volume is between 0.25 and 1.25, and the storage capacity allows for 90 days or more of manure production.[11] We can venture to say that the storage pit used for our study is highly representative of the average manure storing systems in our area.

B. Basic Design Parameters

The following considerations have been taken into account when designing our system:

1.   Materials have to be found without difficulty on the local market, and must be easy to manipulate.
2.   The design has to be simple in order to be readily adaptable to the different pit models in our country.
3.   There has to be a minimum of control and regulation operations, so that the farmer only has to spend a minimum time with the digester.
4.   The low cost of the installations can be paid off within a minimum length of time.

C. Characteristics of the Final Project

They are outlined in Figure 6. The installations will include the following concepts:

Table 4

**TOTAL COSTS PER YEAR, ACCORDING TO AMORTIZATION PERIOD
AND DIFFERENT WORKING HYPOTHESIS (PESETAS)**

Type of financing and manpower use	Number of years of amortization				
	10	5	3	2	1
A + I	270,086,50	349,104,88	462,264,88	606,385,10	1,042,333,31
A + II	220,086,50	299,104,88	412,264,88	556,385,10	992,333,31
B + I	233,425,00	295,849,54	385,245,91	499,100,91	843,499,99
B + II	183,425,00	245,849,54	335,245,90	449,100,91	793,499,99
C + I	182,797,29	222,306,44	278,886,44	350,946,55	568,920,66
C + II	132,797,29	172,306,44	228,886,44	300,946,55	518,920,66
D + I	167,918,00	269,028,00	384,708,00	529,308,00	963,108,00
D + II	117,918,00	219,028,00	334,708,00	479,308,00	913,108,00

A = Bank loan at 16%; B = 21% grant and 79% bank loan at 16%; C = 50% grant and 50% bank loan
at 16%; D = Farmer's money with no interest; I = Maintenance with extra manpower; II = Maintenance
without extra manpower.

- Adapting the pit
- Installation of a gas circuit, including a membrane which will retain the biogas produced
- Installation of a pig excrements circuit

V. ECONOMIC STUDY

This study answers the farmer's most common question when he is trying to reach a
decision concerning the installation of an anaerobic digester: how long will it take for me
to pay off such an installation? The answer to that question depends on several factors.
Some of them are directly imputable to the installation itself, such as its cost, the maintenance
costs, and the biogas produced. Others are related directly to the type of energy used at
present by the farmer for the heating requirements of this operation, e.g., if he is using
expensive electricity, the change to biogas can cut his costs dramatically and the paying off
period will be shorter. Another important factor is the source of the funds insofar as their
cost is concerned. (These installations, in Spain, can receive very favorable official credits
and even subsidies.) We have also taken into account the attitude of the farmer towards
money, in the unlikely but not impossible case of a person who does not think of interest
rates as relevant matters, and perhaps even keeps all his money at home.

We have calculated the total costs per year, considering five different paying off periods:
1, 2, 3, 5, and 10 years, according to different hypothesis on maintenance costs and sources
of the funds (Table 4). As revenues, we have only considered easy-to-quantify parameters,
leaving aside other "revenues" derived, e.g., from the improvement of environmental
conditions (diminution of general contamination, elimination of pathogenic germs, etc.) or
from agronomic consideration (improvement of agronomic virtues of the digested effluent),
or even from the handling conditions (the substrate being easier to manipulate after digestion)
which, although important, are difficult to evaluate. Therefore, as it has been said, we will
only consider as "revenues" the savings in energy that will be achieved by using the biogas
produced instead of a conventional fuel.

A. Revenues

1. Amount of Biogas Produced

Considering that the low-cost digester runs the whole year at psychrophilic temperatures
with HRT = 100 days, the daily production is expected to be situated (according to different

Table 5
COST OF THE THERMIC kWhr FOR DIFFERENT
ENERGY SOURCES

Energy	P.C.I.[a] (kcal/U)	Units	Total thermic efficiency (%)[c]	Cost of the thermic kWhr (pts)
Gas oil	8,122	1	55	8.09
Propane	11,000	kg	60	8.22
Butane	10,900	kg	60	8.99
Electricity	860	kWhr	90	12.18
Biogas	5,500[b]	m³	60	—

[a] P.C.I. = Lower heating power.
[b] Considering 60% CH_4 and 40% CO_2 composition.
[c] Sum of the following efficiencies: boiler + distribution + regulation.

Table 6
ANNUAL REVENUES ACCORDING TO AMOUNT OF
BIOGAS PRODUCTION AND TYPE OF FUEL REPLACED (PTS)

Production hypothesis	Type of energy replaced			
	Diesel	Propane	Butane	Electricity
m³ biogas/ m³ digester-day	(gs)	(pp)	(bt)	(el)
0.1 (x)	113,438.92	115,261.80	126,058.83	170,789.38
0.2 (y)	226,877.84	230,523.60	252,117.66	341,578.76
0.3 (z)	340,316.77	345,785.40	378,176.41	512,368.14

relevant European experiences) between 0.09 and 0.03 m³ of biogas per cubic meter of digester.[5,6,13,14] This wide range of production is due principally to the different chemical composition of the substrate to be digested, and to the temperature conditions of the digester itself throughout the year.

2. Energy Sources Substituted

We consider that these digesters will be installed in farms that already use a conventional fuel for heating purposes. The most commonly used fuels are diesel oil, propane gas, butane gas, and electricity. The characteristics of these fuels are given in Table 5, according to 1984 data.

The average productions that we shall take into account in our study are X = 0.1, Y = 0.2, Z = 0.3 m³ biogas per cubic meter of digester, throughout the year for a 100-m³ digester. In all the cases, the total production equals the net production, since we consider that the energy used by the digester itself, running on a psychrophilic process, is nil.

The amount of biogas produced per year is converted to kilowatt-hours, a "revenue" according to the cost of the kilowatt-hour of the energy source replaced, assuming that the whole biogas production is in effect used (Table 6).

B. Costs

1. Estimated Investment

       1.   Civil works: pit modifications, security valve,   287.000'-ptas
            gasometer support, tubing, etc.

2.   Installations: gas pipes, gasometer, compres-   580.600'-ptas
       sor suppressor, etc.

       $1 = 170$ pesetas (ptas)   Total   867.600'-ptas

2. Sources of Funds

We have taken into consideration four different sources of the funds to pay off this investment:

1.   All the money comes from a private bank loan or from the farmer's own money. Interest rate is 16%.
2.   Some 21% of the investment comes from an official subsidy. The rest (79%) comes from a private bank loan or from the farmer's own money. Interest rate is 16%.
3.   The official subsidy reaches 50% of the investment. The other half comes from a private bank loan or from the farmer's own money. Interest rate is 16%.
4.   The whole investment is paid for with the farmer's own money. No interest considered.

Cases A and B can be considered as the most common ones now. Official subsidies are expected to reach 50% in the future for this kind of installation; this would thus be valid for Case C. In these three cases, amortization annualities are calculated considering compound interest. Case D will provide us with the figures should no interest rates have to be taken into consideration.

3. Maintenance and General Expenses per Year
a. Repairs

The total amount for this chapter is estimated at 7.870 ptas/year.

b. Power consumption

Required only for putting the gas under pressure for heating purposes, and for operating the compressor for mixing the substrate. Estimated at 5,000 ptas/year.

c. Manpower

Here, we consider two hypothesis:

1.   An extra worker will be required 200 hr/year to take charge of the digester. Estimated cost is 50,000 ptas.
2.   The farmer himself assumes the task of caring for the digester, and no extra expenses are necessary.

d. Insurance, Periodic Analysis, and Miscellaneous

We consider for this chapter 3% of the total costs, i.e., 26,028 ptas/year.

4. Amortization

As has been said, we have considered pre-established amortization periods of 1, 2, 3, 5, and 10 years. This, together with the different hypotheses regarding manpower expenses and financing methods, gives the yearly costs reflected in Table 4 (see Figures 11 to 14).

C. Comments
1. Hypthoses A-I and A-II (Figure 11)

The repayment period will be spread over 10 years if the biogas production is low (0.1 m³ biogas per cubic meter of digester per day). If it were medium (0.2 m³ biogas per cubic meter of biogas per day) the manpower factor is crucial to reduce the repayment period.

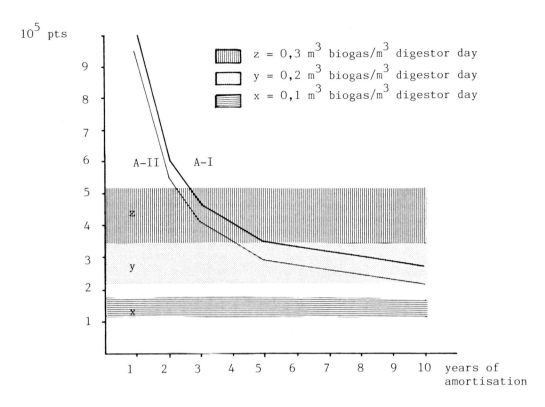

FIGURE 11. Annual cost according to amortization periods, Cases A-I and A-II, and revenues according to type of fuel replaced and amount of biogas produced.

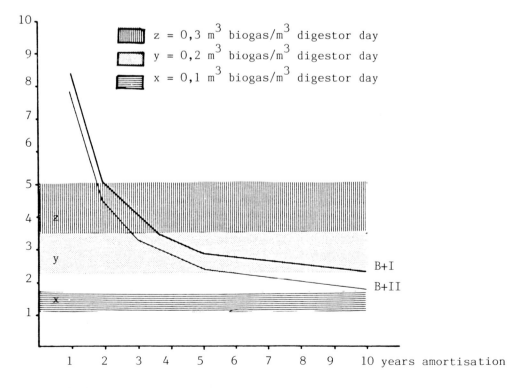

FIGURE 12. Annual cost according to amortization periods, Cases B-I and B-II, and revenues according to type of fuel replaced and amount of biogas produced.

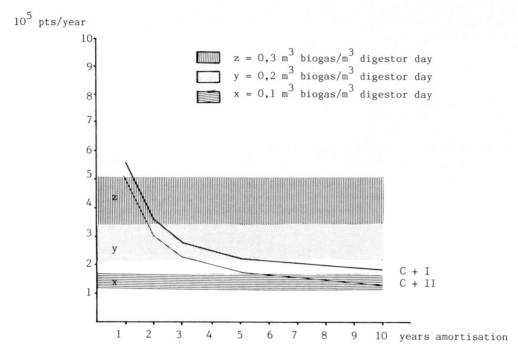

FIGURE 13.   Annual cost according to amortization periods, Cases C-I and C-II, and revenues according to type of fuel replaced and amount of biogas produced.

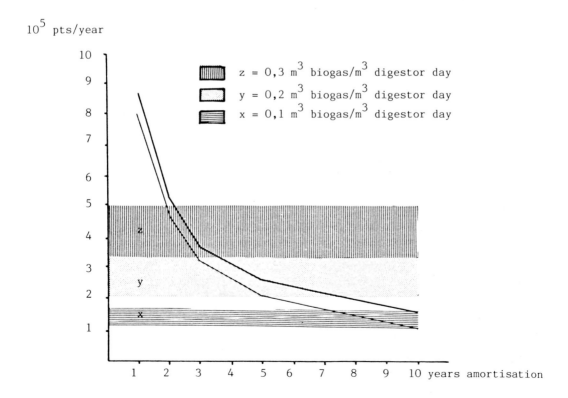

FIGURE 14.   Annual cost according to amortization periods, Cases D-I and D-II, and revenues according to type of fuel replaced and amount of biogas produced.

With a high production (0.3 m³ of biogas per cubic meter of digester per day), the amortization period will be less than 5 years.

2. Hypotheses B-I and B-II (Figure 12)
With low production, the repayment periods are also longer than 10 years. With medium production, they will be somewhere between 4 and 10 years. High production reduces this period to 2 to 4 years.

3. Hypotheses G-I and G-II (Figure 13)
A 50% subsidy and a bank loan at 16% for the remainder, both with and without manpower. When biogas production is low, the amortization period will be 5 to 10 years, depending upon the type of energy replaced, only when outside manpower does not apply. With medium production, the amortization period is shorter than in Cases A and B, i.e., less than 5 years, even including outside manpower, and it would be even less than 2 years when gas production is high.

4. Hypotheses D-I and D-II (Figure 14)
Financing with the farmer's own money, both with and without outside manpower. Without manpower, unlike Cases A and B, when the biogas production is low, the installation can be paid off in less than 10 years. With manpower, the amortization period will be 10 years only when the type of energy replaced is electricity. With medium production, the situation is similar to Case C. On the contrary, with high production, the situation becomes comparable to case B.

VI. GENERAL CONCLUSIONS

A. Temperature Study of the Pit
The daily temperature oscillation of the pit was 1°C. At the same time, the temperature was never less than 10°C. These two factors permit a psychrophilic anaerobic fermentation throughout the year. During the summer period, the temperatures were more than 20°C, permitting a mesophilic fermentation throughout the year. During the summer period, the temperatures were more than 20°C, permitting a mesophilic fermentation with higher gas production.

It became necessary to continue this temperature study with a second phase consisting in an in-depth study of the apport and heat losses. The final objective would be to make a mathematical model of the temperature behavior of the pit.

B. Digester Design and Economic Study
The proposed design for the conversion of an existing pit into an anaerobic digester can be adapted to virtually any one of the reservoirs currently in use. What is more, the materials used for the operation can be found without difficulty on the local markets, and can be easily installed and managed. The cost of such an installation could be around 9000 ptas per cubic meter of digester (those installed up to now are 5 to 20 times more expensive; (see Figure 5).

The range of revenues, taking into account the different biogas production levels and several types of energies replaced, is very wide, with such extreme values of ptas 113.483,92 per year to ptas 512.368,14 per year.

The influence of the cost of manpower is small considering short amortization periods, but it is basic with long amortization periods. One of the main factors for achieving short amortization periods is the amount of biogas production. With high production levels, the amortization period is always less than 5 years. With intermediate production, the amortization periods range from 2 to 10 years. With low production, the amortization period of the installation will be less than 10 years in only two cases:

1. When there are no outside manpower costs, or when there is a subsidy of 50% of the total costs.
2. When installation is paid for with the farmer's own money, and no interest rates have been considered.

If the prices of conventional energy sources were to increase faster than the overall costs, then the amortization periods would be proportionately shorter.

In the low-cost anaerobic digestion plant now under construction in Caldes de Montbui (Barcelona), the following parameters will be studied over a 3-year period: total gas production, degree of purification obtained, maintenance, and management problems. This will allow us to obtain the real costs of the operation, and to determine the actual amortization periods.

REFERENCES

1. **Hobson, P. N., Bousfield, S., and Summers, R.,** *Methane Production from Agricultural and Domestic Wastes,* Applied Science, London, 1981, 235.
2. U.N. Guidebook on Biogas Development, *Energy Resources Development Series, United Nations,* New York, 1980, 127.
3. **Van Buren, A.,** *A Chinese Biogas Manual,* Intermediate Technology Publ., London, 1979, 136.
4. Government of Taiwan, *Inform Taiwan Provincial Livestock Research Institute, Taiwan,* Livestock Waste Disposal Experiment Center, Taiwan, 1981, 30.
5. **Gkopiste, M. and Eggersolub, P.,** Psycophile. Digestion systems impractice, *Biomass,* Berlin, Germany, 1982.
6. **Wellweger, A. and Kaugmann, R.,** Production de biogas dans les installations no changées a partir de lisier porcin, *Document. Tech. Agric.,* 2, 150, 1982.
7. **Wellinger, A., Fisher, R., and Benson, J. M.,** Production de biogaz a partir de lisier bovin et porcin dans des explotacions agricoles suisses, *Trib. Cebedeau,* 455, 2, 429, 1981.
8. **Meynell, P. J.,** *Methane Planning a Digester,* Prims Press, London, 1982, 162.
9. **Van Velsen, A. F. M.,** *Anaerobia Digestion of Piggery Waste,* Netherland Agricultural University, Wegeningen, 1981, 103.
10. **Welinger, A.,** Les parametres inflençant la digestion metagene, *Biogaz Suisse,* 1982, 70.
11. **Ferrer, P., Sans, J. B., and Pomar, J.,** Utilización agrícola del estiércol líquido do porcino, *Fulls d'Informació Tècnica. J. Servei d'Extensió Agrària,* 14, 1981.
12. **Ossombo, N.,** Fermentation methanique anaerobie d'Ecichhordia Crapssipes (Jacinthe d'Eau), Etude des Conditions de Mise en Fermentation, Influence de quelques Pàrametres Physicochimiques, Ph.D. thesis, University of Paris II, Paris, 1983.
13. **Kaugmann, R., Sutter, K., and Welinger, A.,** Low temperature methane generation from animal manure. Energy conservation and use of renouvable Energies in the *Bic-Industries* 2, Pergamon, Press, Oxford, 1982.
14. **Sangiorgio, F., Balsareny, P., and Bonfanty, P.,** *Low Cost Biogas Installations in Italian Agriculture,* Instituto di Ingegneria agraria, Milano, 1983.
15. **Meseguer, J.,** personal communication, 1984.

Chapter 6

BIOFILM KINETICS — MASS TRANSFER EFFECTS AND THEIR IMPLICATION TO PROCESS DESIGN, OPERATION, AND CONTROL

Wen K. Shieh

TABLE OF CONTENTS

I. INTRODUCTION

Biological treatment is probably the most important step in processing municipal and industrial wastewaters. Biological treatment systems are living systems which depend upon mixed microbial population to remove the nonsettleable colloid solids and to stabilize the organic matter. With proper analysis and environmental control, almost all wastewaters can be treated biologically.[1]

The biological wastewater treatment systems currently employed can be categorized into two types according to the physical states of the microbial population involved. The first type is called the suspended-growth systems in which the microorganisms reponsible for the removal of the organic matter or other constitutents are suspended in the liquid. The suspension of microorganisms within the liquid is accomplished by using diffused air or mechanical aeration. In order to keep the microorganisms within the system for a sufficient period of time, recirculation of settled microbial solids from the effluent via a final clarifier is employed. The activated sludge systems are prominent examples.

The second type of biological wastewater treatment systems are called the attached-growth systems. In these systems the microorganisms are attached to some inert media, such as rocks, slags, sand, and plastic materials. The formation of biofilms on the media surface immobilizes most of the microbial population within the system. Therefore, recirculation of settled microbial solids is not employed in the attached-growth systems. The attached-growth systems are often referred to as the biofilm systems.

A. Biofilm Systems Used in Wastewater Treatment

The following four types of biofilm systems have been employed in wastewater treatment: trickling filters, rotating biological contactors, submerged filters, and fluidized bed biofilm reactors. Figure 1 is a schematic presentation of these systems.

Among these biofilm systems, trickling filters are the oldest; in fact, the first trickling filters with distribution of the influent over the biofilm surface were built in 1981 at the Lawrence Experiment Station in Massachusetts, and in 1893 at Salford, England.[2] The highly permeable plastic media are used in the modern trickling filter installations to increase the hydraulic and organic loadings applied. Many variations have been proposed but the basic design principles are the same.

The rotating biological contactor (RBC), like the trickling filter, is a three-phase system but the media move through the liquid and gas phases rather than being fixed in space. Alternate exposure of the biofilms to liquid and air during rotation allows adequate supply of the organic matter and the oxygen for the microbial population retained within the biofilms. The biofilm thickness is controllable by rotation.

The use of RBC technology in Europe has been extensive, and over 700 installations are reported in operation.[3] The number of RBC facilities in the U.S. has been significantly increased since 1972.[3]

The submerged filters are two-phase systems. For aerobic applications, oxygen has to be dissolved in the wastewater before entering the filter. Utilization of small media in submerged filters may provide filtration in addition to the biochemical reactions performed. The submerged filters have to be backwashed periodically in order to prevent filter clogging because accumulation of biofilms is uncontrollable. Two types of submerged filters are used: downflow and upflow.

Submerged filters are widely employed in the anaerobic treatment of concentrated wastewaters.[4-11] They have also been applied in biological denitrification.[12-15]

The fluidized bed biofilm reactor (FBBR) is a new adaptation of biofilm systems in which the media are fluidized in the wastewater to be treated. The fluidization of the small media (e.g., sands) results in a large surface area for microbial attachment and growth. As a result,

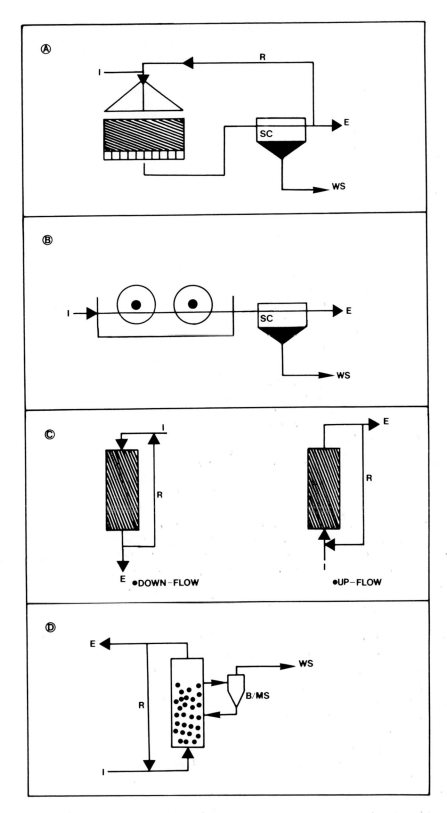

FIGURE 1. Biofilm systems used for wastewater treatment. (A) Trickling filter; (B) rotating biological contactor; (C) submerged filter; (D) fluidized bed biofilm reaction. I: Influent; E: effluent; R: recycle; WS: waste sludge; SC: secondary clarifier; B/MS: biomass/media separator.

Table 1
BIOFILM SYSTEMS VS. SUSPENDED-GROWTH SYSTEMS IN WASTEWATER TREATMENT

Biofilm systems	Suspended-growth systems
Microorganisms are retained within the biofilms attached to the media.	Microorganisms are suspended in the wastewater by means of mixing and/or aeration.
Fair protection against the adverse effects of toxicants in the influent.	Sensitive to the adverse effects of toxicants in the influent.
Inter- and intra-phase mass transfer is significant because the biofilm structure tends to retard the rate of transport of substrate through it. Liquid-biofilm interface forms another resistance to the transport of substrate across it. Therefore, the biofilm systems are heterogeneous systems.	Inter- and intra-phase mass transfer is insignificant. Vigorous mixing and/or aeration reduces the thickness of the liquid film surrounding the flocs and the sizes of the flocs which, in turn, reduce the effects of mass transfer. Therefore, the suspended-growth systems are treated as the homogeneous systems.
The system performance is not affected by the performance of the secondary clarifier. The biomass concentration in the reactor is maintained through the attachment of microorganisms on the media surface and the subsequent growth of biofilms.	The system performance is intimately linked to the performance of the secondary clarifier because the maintenance of the desirable biomass concentration in the reactor depends upon the recirculation of concentrated microbial solids from the secondary clarifier with proper thickening function.
Secondary clarification in some cases may be eliminated because the suspended solids level in the system effluent is very low. This is possible because most of the biomass is retained within the biofilms rather than suspended in the wastewater.	Secondary clarification is required to reduce the effluent suspended solids concentration to the acceptable level.
Unified design approaches are not fully developed yet because the biofilm systems are generally complex.	Unified design approaches such as F/M approach and SRT approach are now well developed and widely employed for practical applications.

high biomass concentration can be maintained which, in turn, reduces the required reactor volume or treatment time. Another benefit is that the fluidization of the media eliminates the problems of high head loss and bed clogging associated with the operation of submerged filters. Biofilms in an FBBR may be controlled either by a device for shearing the biofilms off the media at the top of the reactor or by withdrawal of the bioparticles (biofilm coated media) for biofilm removal and recirculation of cleaned media.[16]

B. Biofilm Systems Vs. Suspended-Growth Systems in Wastewater Treatment

A comparison of the characteristics of biofilm systems to those of suspended-growth systems in wastewater treatment is summarized in Table 1. Two distinct differences between these two systems deserve further discussion.

First, the inter- and intra-phase mass transfer phenomena are significant in the biofilm systems because the biofilm structure tends to retard the rate of transport of substrate through it. On the other hand, such resistances are insignificant in the suspended-growth systems because the vigorous mixing provided significantly reduces the mass transfer limitations.[17]

Second, as far as the design practice is concerned, no unified approaches are currently available for the biofilm systems used in wastewater treatment. For instance, more than 15 models have been proposed for the design of trickling filters, each of them is applicable only under rather limited conditions.[18] On the other hand, unified design approaches such as F/M and solid retention time (SRT) techniques are well established in the design practice of the suspended-growth systems used for wastewater treatment. These approaches are now commonly used for the design of full-scale suspended-growth wastewater treatment facilities.[1]

C. Biofilm Kinetics

Recent advents in biofilm system engineering as well as improved knowledge about the fundamental reaction phenomena of the biofilm conversion process have promoted a resurgence of interest in the development of mechanistic biofilm models.[19,20] Although these mechanistic models are rather complex as compared to the empirical ones, their applications in real-world situations can be facilitated with modern-day computer technology.

An adequate model of the reactions taking place in a biofilm system should provide two types of information: (1) a kinetic description of the path followed by the substrate from the bulk of the liquid until it is consumed at the active biofilm sites and (2) a theoretical basis which allows, at least in principle, the computation of the parameters included in the kinetic description from the physicochemical properties of the system.[21,22]

An attempt is made in this chapter to elucidate and analyze those factors which may have a significant impact on the performance of a biofilm system. Analysis is made possible by the definition of the effectiveness factor, which is a simple means to define the kinetic regime of a biofilm system. Techniques commonly used in modeling heterogeneous catalytic systems are used herein to gain better insight into the basic events occurring within a biofilm system.

II. THEORETICAL CONSIDERATIONS

Because virtually all the biomass in a biofilm system is part of the biofilm, substrates must be transported into the biofilm before they can be metabolized. In a simple but general sense, three mass transfer and reaction steps are involved in the overall reaction of a biofilm system.[22] They are (1) substrate transport from the bulk liquid to the biofilm surface through a liquid film surrounding the biofilm (external mass transfer), (2) substrate transport in the biofilm (internal mass transfer), and (3) substrate conversion in the biofilm (step 3). Steps 2 and 3 occur simultaneously; thus, they will have a single rate. Step 1, and the overall internal mass transfer-and-substrate conversion phenomenon (steps 2 and 3) occur in series; thus the slowest step will become the rate limiting one in these sequential steps. Therefore, it is essential that the combined effect of microbial rate processes and physical mass transfer phenomena are considered when modeling biofilm systems.

Another important consideration in modeling biofilm systems is the estimation of biomass concentration. Because the substrate conversion rates in the biofilm systems are a function of biomass present, these rates may be calculated using media surface area, biofilm density, and biofilm thickness.[23] Unlike conventional biofilm systems, the media surface area in an FBBR is a direct function of the degree of fluidization maintained in the reactor. Therefore, an analysis of fluidization mechanics within a FBBR is required to yield two critical parameters for the estimation of biomass concentration: equilibrium biofilm thickness and bed porosity.[24]

A. Mass Transfer Limitations — Concept for Effectiveness Factor

The gelatinous biofilm tends to retard substrate transport and thus cause the substrate concentration surrounding the microorganisms to be less than that in the bulk liquid. For intrinsic reaction rates with positive dependence on the substrate concentration (i.e., first order, Michaelis-Menten, etc.) the gradient established by steps 1 and 2 decreases the observed reaction rate by decreasing local (i.e., intrabiofilm) substrate concentration. For intrinsic zero-order kinetics, steps 1 and 2 can decrease the observed reaction rate by limiting the substrate penetration depth into the biofilm.[25]

It has been demonstrated that external mass transfer effects can be neglected with acceptable magnitude of errors when the kinetics of the biofilm systems are assessed.[26,27] This greatly simplifies the mathematics which result. Therefore, the analysis of biofilm kinetics described herein will proceed with consideration of only simultaneous intrabiofilm mass transfer and reaction.

1. Effectiveness Factor Expressions

The effect of mass transfer on the intrinsic reaction rate is commonly quantified through the definition of an effectiveness factor.[28] Under the assumption of negligible external mass transfer, the effectiveness factor (η) is defined as:

$$\eta = \frac{\text{observed reaction rate}}{\text{intrinsic reaction rate at bulk-liquid conditions}} \tag{1}$$

Mathematical expressions for the biofilm effectiveness factor will be developed for both planar and spherical media. The former is applicable in tricklng filters, RBCs, and submerged filters whereas the latter is applicable in FBBRs. The following assumptions are made in the analysis of biofilm kinetics:

1. Homogeneous biofilm of uniform thickness
2. Uniform media dimensions
3. Internal mass transfer described by Fick's first law
4. Single limiting soluble substrate
5. Intrinsic zero-order kinetics
6. Completely mixed conditions in bulk liquid
7. Steady-state conditions

The biofilm continuity equations for the limiting substrate are then:

$$D_e \frac{d^2S}{dx^2} = \rho k_o \text{ (planar media)} \tag{2}$$

$$\frac{D_e}{r^2} \frac{d}{dr}\left(r^2 \frac{dS}{dr}\right) = \rho k_o \text{ (spherical media)} \tag{3}$$

with the following boundary conditions apply (Figure 2):

$$S = S_b \qquad \text{at}\begin{cases} x = 0 \text{ (planar media)} \\ r = r_p \text{ (spherical media)} \end{cases} \tag{4}$$

$$S = \frac{dS}{dx} = 0 \quad \text{at } x = x_c \text{ (planar media)} \tag{5}$$

$$S = \frac{dS}{dr} = 0 \quad \text{at } r = r_c \text{ (spherical media)} \tag{6}$$

where D_e is the effective diffusivity of the substrate in the biofilm, L^2/T; x is the distance measured from the biofilm surface, L; S is the substrate concentration in the biofilm, M/L^3; ρ is the biofilm dry density, M/L^3; k_o is the intrinsic zero-order rate constant, $M/M - T$; r is the radial distance measured from the bioparticle center, L; S_b is the bulk-liquid

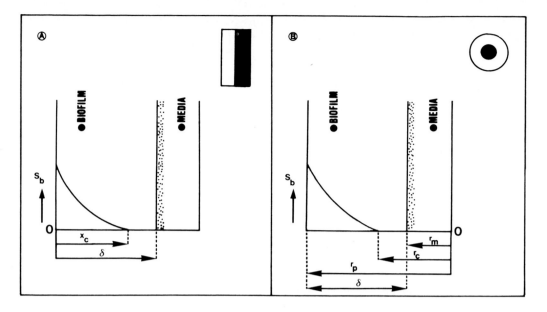

FIGURE 2.   Boundary conditions for intrinsic zero-order biofilm systems. (A) Planar media; (B) spherical media.

substrate concentration, M/L^3; r_p is the bioparticle radius, L; x_c and r_c are the substrate penetration depths for planar and spherical media, respectively, L.

Integration of Equations 2 and 3 subject to the boundary conditions defined in Equations 4 to 6 yields:

$$x_c = \left\{ \frac{2D_eS_b}{\rho k_o} \right\}^{0.5} \tag{7}$$

$$\left\lfloor \frac{r_c}{r_p} \right\rfloor^3 - 1.5 \left\lfloor \frac{r_c}{r_p} \right\rfloor^2 + \left\lfloor 0.5 - \frac{3}{\phi_o^2} \right\rfloor = 0 \tag{8}$$

where $\phi_o = (\rho k_o/D_eS_b)^{0.5}$, the conventional zero-order Thiele modulus.[28]

The effectiveness factor for the intrinsic zero-order reaction (η_o) is simply the ratio of biofilm volume with substrate concentration greater than zero to the total biofilm volume.[29] Thus,

$$\eta_o = \frac{\left\{ \frac{2D_eS_b}{\rho k_o} \right\}^{0.5}}{\delta} \quad \text{(planar media)} \tag{9}$$

$$\eta_o = \frac{1 - (r_c/r_p)^3}{1 - (r_m/r_p)^3} \quad \text{(spherical media)} \tag{10}$$

where δ is the biofilm thickness, L, and r_m is the media radius, L.

Numerical solution of Equations 8 and 10 yields an explicit relationship between η_o and S_b in partially penetrated bioparticles:[29]

$$\eta_o = \frac{3.184}{1 - \left[\frac{r_m}{r_p} \right]^3} \left[\frac{D_e}{\rho k_o r_p^2} \right]^{0.5} S_b^{0.5} \tag{11}$$

Both Equations 9 and 11 indicate that, for the intrinsic zero-order reaction in partially penetrated biofilms, the effectiveness factor is proportional to the bulk-liquid substrate concentration to the 0.5 power. The difference in media geometry is irrelevant.

2. Overall Rates of Substrate Conversion in Biofilm Systems

Because recirculation of reactor effluent is commonly employed in the operation of the biofilm systems in order to insure adequate organic and hydraulic loadings, the overall rate expressions for the substrate conversion are strongly affected by the hydraulic characteristics of the systems. It has been reported elsewhere that completely mixed biofilm systems are generally used in practice for wastewater treatment.[30-33] Therefore, the theoretical analysis will proceed based on the assumption of completely mixed conditions.

A mass balance on the bulk-liquid substrate concentration (S_b) across a biofilm system yields:

$$Q(S_{bi} - S_b) = VR_v \tag{12}$$

where Q is the wastewater flowrate, L^3/T; S_{bi} is the influent substrate concentration. M/L^3; V is the reactor volume, L^3; and R_v is the observed substrate conversion rate per unit reactor volume, $M/L^3 - T$.

The reaction term (R_v) in Equation 12 is more useful when expressed as:

$$R_v = RX \tag{13}$$

where R is the observed substrate conversion rate per unit biofilm mass, $M/M - T$; X is the biomass concentration in the reactor, M/L^3.

The observed substrate conversion rate per unit biofilm mass (R) is related to the effectiveness factor (η_o) by:[29]

$$R = \eta_o k_o \tag{14}$$

Thus, for planar media the observed substrate conversion rate in the reactor is

$$\frac{(S_{bi} - S_b)}{\theta} = (2\rho k_o D_e)^{0.5} a S_b^{0.5} \tag{15}$$

where θ is the hydraulic retention time (HRT), T; a is the specific surface area of the media, $1/L$. For spherical media, then:

$$\frac{(S_{bi} - S_b)}{\theta} = \frac{3.184}{1 - \left[\dfrac{r_m}{r_p}\right]^3} \left[\frac{D_e}{\rho k_o r_p^2}\right]^{0.5} X S_b^{0.5} \tag{16}$$

Therefore, the observed substrate conversion rate in an intrinsic zero-order biofilm system limited by internal mass transfer is described by a half-order equation. This is an interesting result because of the alternation of the intrinsic (true) reaction order. This demonstrates that a mechanisitic biofilm model is capable of explicitly expressing the influence and interrelationships of individual reaction phenomena involved in the system. The empirical biofilm models are far less useful in this aspect.

Equation 15 indicates that the observed substrate conversion rate in a conventional biofilm system is directly proportional to the specific surface area provided by the media used. Because the media within the conventional biofilm systems are fixed in space either by gravity or by direct attachment to the reactor wall, the hydraulic characteristics of the systems will not affect the specific surface area of the media. Nevertheless, the situation is completely different in the FBBRs where the media are retained in suspension by drag forces exerted by the upflowing wastewater. Therefore, one would expect that the biomass concentration term (X) in Equation 16 is a variable depending on the hydrodynamic environment prevailing in an FBBR. Further analysis is presented as follows.

3. Fluidization Mechanics in the FBBRs

The main difference between conventional biofilm systems and FBBRs is the free movement of bioparticles in the latter. Moreover, the growth of biofilm changes the overall density of the bioparticles and, therefore, the expansion of the fluidized bed. Therefore, the number and size of bioparticles (and thus, specific surface area and biofilm thickness) per unit fluidized bed volume are two pieces of critical information for the estimation of FBBR biomass concentration. Analysis of the fluidization mechanics in an FBBR provides a logical and convenient way to obtain this information.[34-36]

Several solid-liquid fluidization correlations, which link particle concentration in the fluidized state to the physical characteristics of a fluidized bed system which are measurable, are available.[37-40] Among them the Richardson-Zaki correlation is widely used:

$$\frac{U}{U_t} = \epsilon^n \tag{17}$$

where U is the superficial upflow velocity of the wastewater through an FBBR, L/T; U_t is the bioparticle terminal settling velocity, L/T; ϵ is the bed porosity; n is the expansion index.

The following two empirical correlations have been proposed by Shieh and Chen to relate both U_t and n to the Galileo number (N_{Ga}) which defines the physical characteristics of an FBBR:[35]

$$U_t = 5753.71 N_{Ga}^{-0.8222} \tag{18}$$

$$n = 47.36 N_{Ga}^{-0.2576} \tag{19}$$

and

$$N_{Ga} = \frac{8r_p^3 (\rho_p - \rho_\ell) \rho_\ell g}{\mu^2} \tag{20}$$

$$\rho_p = \rho_m \left[\frac{r_m}{r_p}\right]^3 + \frac{\rho}{(1 - P)} \left\{1 - \left[\frac{r_m}{r_p}\right]^3\right\} \tag{21}$$

where ρ_p is the bioparticle density, M/L³; ρ_ℓ is the wastewater density, M/L³; g is the gravitational acceleration, L/T²; μ is the wastewater dynamic viscosity, M/T − L; ρ_m is the media density, M/L³; P is biofilm moisture content.

Using Equations 17 to 21, the biomass concentration in an FBBR is calculated as:

$$X = \rho(1 - \epsilon)\left\{1 - \left[\frac{r_m}{r_p}\right]^3\right\} \tag{22}$$

Prediction of biomass concentration in an FBBR utilizing the correlations described herein can be facilitated via the following interactive procedure developed by Mulcahy and LaMotta:[41]

1.   Guess a biofilm thickness, δ
2.   Calculate the bioparticle radius:

$$r_p = r_m + \delta \tag{23}$$

3.   Calculate the bioparticle density using Equation 21
4.   Calculate the Galileo number using Equation 20
5.   Calculate the bioparticle terminal settling velocity using Equation 18
6.   Calculate the expansion index using Equation 19
7.   Specify a sperficial upflow velocity, U
8.   Calculate the resultant equilibrium bed porosity using Equation 17
9.   Specify the total volume of media used in the reactor, V_m
10.   Calculate the resultant trial expanded fluidized bed height, $\hat{H}_B$, using a volume balance on solids within the reactor:

$$\hat{H}_B = \frac{V_m}{A(1 - \epsilon)} \left[\frac{r_m}{r_p} \right]^3 \tag{24}$$

where A is the cross-sectional area of the reactor, L^2

11.   Compare the calculated expanded fluidized bed height ($\hat{H}_B$) with the prescribed expanded fluidized bed height (H_B)
12.   If unacceptable, return to step 1 and repeat the procedure for a new δ
13.   If acceptable, use Equation 22 to calculate the biomass concentration

III. SENSITIVITY ANALYSIS

A sensitivity analysis utilizing the biofilm models developed is performed in this section to elucidate those parameters which may have a significant impact on the design, operation, and control of a biofilm system.

As discussed previously, the effectiveness factor is a convenient means to quantify the effects of mass transfer and to define the kinetic regime of a biofilm system. According to Equations 9 and 11, the zero-order biofilm effectiveness factor (η_o) is a function of physical and microbiological properties of the system. The number of variables in Equations 9 and 11 can be reduced by incorporating them into a dimensionless parameter as defined below:

$$\phi_{om} = \bar{\delta} \left[\frac{\rho k_o}{D_e S_b} \right]^{0.5}, \text{ a modified zero-order Thiele modulus} \tag{25}$$

where $\bar{\delta}$ is the characteristic biofilm thickness which is defined as:

$$\bar{\delta} = \frac{\text{biofilm volume}}{\text{biofilm exterior surface area}} \tag{26}$$

Therefore:

$$\bar{\delta} = \delta \text{ (planar media)} \tag{27}$$

$$\bar{\delta} = \frac{r_p^3 - r_m^3}{3r_p^2} \text{ (spherical media)} \tag{28}$$

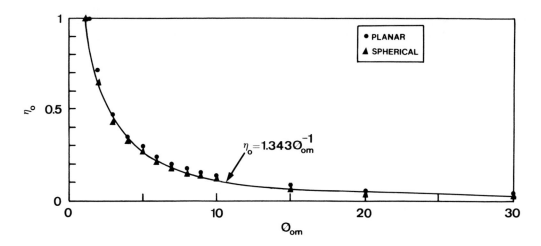

FIGURE 3. Zero-order biofilm effectiveness factor (η_o) as a function of modified zero-order Thiele modulus (ϕ_{om}).

The functional relationship between η_o and ϕ_{om}, which can be obtained through numerical solution of Equations 9, 11, 25, 27, and 28 is illustrated in Figure 3. This functional relationship is well described by the following empirical equation:

$$\eta_o = 1.343\phi_{om}^{-1} \qquad (29)$$

An explicit relationship is thus established for the intrinsic zero-order reaction in partially penetrated biofilms which shows the effectiveness factor to be inversely proportional to the modified zero-order Thiele modulus. Setting $\eta_o = 1.0$ in Equation 29 gives $\phi_{om} = 1.343$, the value at which transition occurs from full to partial substrate penetration of the biofilm. For $\phi_{om} \leq 1.343$, the whole biofilm is active and intrinsic zero-order kinetics are observed. For $\phi_{om} > 1.343$, the inner portion of the biofilm is substrate starved and the observed reaction rate is proportional to $S_b^{0.5}$.

A. Limiting Biofilm Thickness for Full Substrate Penetration

For a biofilm system designed for a specific application, the limiting biofilm thickness for full substrate penetration (δ_ℓ) can be calculated by setting $\eta_o = 1.0$ in Equation 29 and solving for δ. Thus:

$$\delta_\ell = 1.343 \left[\frac{D_e}{\rho k_o}\right]^{0.5} S_b^{0.5} \text{ (planar media)} \qquad (30)$$

$$\delta_\ell^3 + 3\delta_\ell^2 \left\{r_m - 1.343\left[\frac{D_e S_b}{\rho k_o}\right]^{0.5}\right\} + 3\delta_\ell\left\{r_m^2 - 2.686\left[\frac{D_e S_b}{\rho k_o}\right]^{0.5}\right\}$$

$$-4.029\left[\frac{D_e S_b r_m^4}{\rho k_o}\right]^{0.5} = 0 \text{ (spherical media)} \qquad (31)$$

Equations 30 and 31 can be used to calculate δ_ℓ values that are likely to be expected in both conventional biofilm systems and FBBRs for a number of intrinsic zero-order reactions commonly observed in biological wastewater treatment. The results are illustrated in Figure

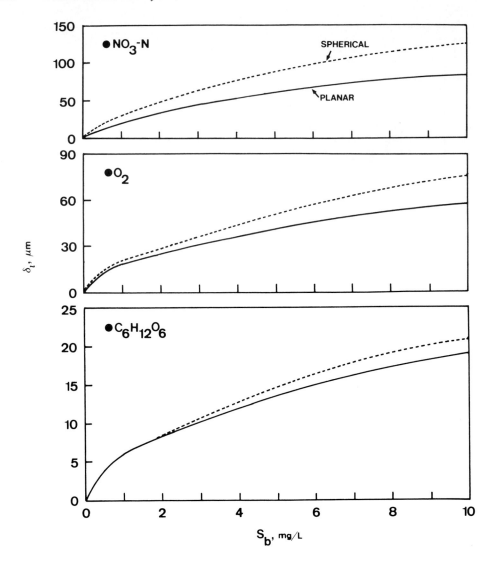

FIGURE 4.   Limiting biofilm thickness for full substrate penetration (δ_ℓ) as a function of bulk-liquid substrate concentration (S_b). Radius of the spherical media used (r_m): 0.2 mm.

4. It is seen from Figure 4 that higher bulk-liquid substrate concentration is required to penetrate into thick biofilm to insure that a majority of the biofilm is active. For instance, in a denitrification FBBR, complete penetration of a 100-μm biofilm grown on 0.4-mm diameter media will occur when the bulk-liquid NO_3-N concentration is equal or greater than 6 mg/ℓ. For NO_3-N values less than 6 mg/ℓ only partial nitrate penetration will occur. Experimental evidence indicates that biofilms with a thickness exceeding 150 μm are commonly found in the biofilm systems designed for denitrification.[42-45] Therefore, the observed denitrification rates in the biofilm systems are likely to be internal mass transfer limited for NO_3-N levels commonly found in nitrified municipal wastewaters.

For the biofilm systems designed for the aerobic treatment of wastewater, dissolved oxygen is likely to become a rate-limiting factor because the dissolved oxygen levels maintained in the reactors are generally around 2 mg/ℓ. Under these conditions, a δ_ℓ value less than 30 μm is required in order to insure that the whole biofilm is aerobic. The equilibrium biofilm thickness in a full-scale, aerobic biofilm system is most likely well above this limiting value because of the fast growth rates of the heterogeneous microorganisms.

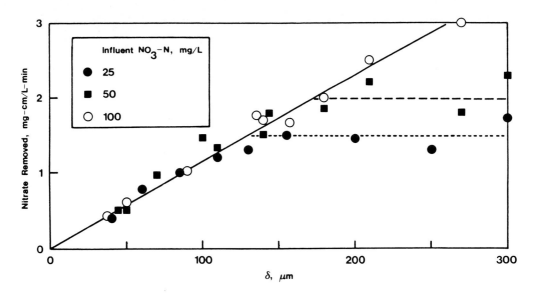

FIGURE 5. The observed denitrification rate in a rotating-disk biofilm system as a function of biofilm thickness (δ).[46]

Although the total mass of biofilms in the conventional biofilm systems is a direct function of biofilm thickness, it is important to realize that a thicker biofilm does not necessarily have a greater substrate conversion rate than a thin biofilm because of the mass transfer resistances imposed by the gelatinous structure of the biofilm. This conclusion is substantiated by the experimental evidence reported by Mulcahy et al. which is illustrated in Figure 5.[46] As shown, the observed denitrification rate in a rotating-disk biofilm system increases linearly with biofilm thickness up to a certain thickness, beyond which it remains constant. When the biofilm thickness is less than a critical value (i.e., δ_ℓ), the whole biofilm is active. Therefore, the observed denitrification rate is directly proportional to the amount of active biomass in the biofilm, and thus the biofilm thickness. As the biofilm thickness exceeds the critical value, only that portion of the biofilm with thickness less than the critical value is still active. As a result, further increase in total biofilm thickness will not induce a corresponding increase in the observed denitrification rate.

The active biomass concentration in a conventional biofilm system can be calculated through the definition of effectiveness factor. The total biomass concentration (X) is

$$X = \rho \delta a V \tag{32}$$

and the active biomass concentration (X_a) is

$$X_a = \rho x_c a V \tag{33}$$

From Equations 7 and 9 it is clear that

$$\eta_o = \frac{x_c}{\delta} \tag{34}$$

Therefore:

$$X_a = \eta_o X \tag{35}$$

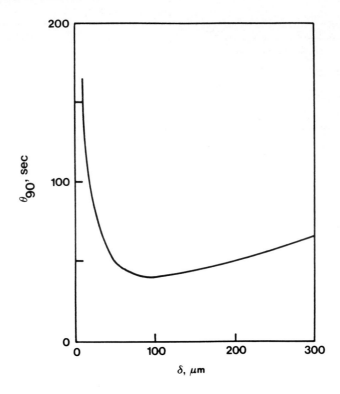

FIGURE 6.   Effect of biofilm thickness (δ) on the required HRT for
90% removal of NO_3-N (θ_{90}). $S_{bi} = 20$ mg/ℓ; $r_m = 0.22$ mm; $\rho_m = 2.65$ g/mℓ; $\rho = 65$ mg/mℓ; U = 1 cm/sec; P = 0.93; $k_o - 3.32 \times 10^{-5}\ell$/sec, and $D_e = 0.815 \times 10^{-5}$ cm^2/sec.

The concept of active biomass concentration is equally applicable in the FBBRs. Nevertheless, the situation is somewhat different in an FBBR where the active biomass concentration present is a function of both the biofilm thickness and the degree of fluidization prevailing in the system. If an FBBR is operated in the hydraulic regime where the rate of increase in active biomass due to biofilm accumulation is faster than the rate of decrease in active biomass due to the reduction in bioparticles per unit fluidized bed volume caused by bed expansion, then the observed substrate conversion rate will increase with biofilm thickness. Otherwise, the opposite holds. Therefore, under a given set of operating conditions, there always exists a biofilm thickness which yields the maximum active biomass concentration in the reactor (i.e., maximum $\eta_o X$ or X_a) and thus, the optimal substrate conversion rate. The evidence of this may be seen in Figure 6 in which the effect of biofilm thickness on the required HRT for 90% NO_3-N removal in an FBBR (θ_{90}) is plotted.

Therefore, the rate of reaction in an FBBR is directly proportional to the active biomass concentration in the reactor. Maximization of the active biomass concentration will therefore maximize the overall rate of substrate conversion in the FBBR. Optimization of FBBR performance is, nevertheless, not a straightforward process because of the effect of fluidization on the effectiveness factor and biomass concentration. Further discussion on the fluidization mechanics in an FBBR is required.

B. Fluidization Mechanics in an FBBR
Equation 32 indicates that the biomass concentration in a conventional biofilm system is directly proportional to biofilm thickness for a given type of media used. Nevertheless, the biomass concentration in an FBBR is dependent on a number of interrelated factors in

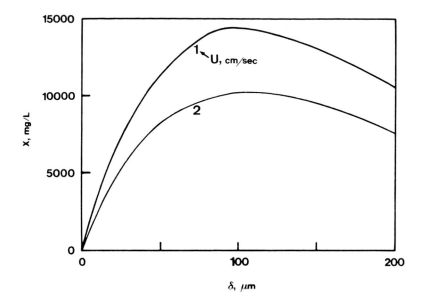

FIGURE 7. Effect of biofilm thickness (δ) on the biomass concentration (X) in an FBBR. $r_m = 0.2$ mm; $\rho_m = 2.65$ g/mℓ; $\rho = 65$ mg/mℓ, and P = 0.93.

addition to biofilm thickness because of the fluidization state maintained in the system. Mulcahy and LaMotta have shown that the equilibrium biofilm thickness in an FBBR is dependent on superficial upflow velocity, expanded fluidized bed height, media size and density, and media volume.[41] As background to an understanding of this phenomenon, the relationship between biomass concentration and equilibrium biofilm thickness is determined via the iterative procedure as developed by Mulcahy and LaMotta.[41] The results are illustrated in Figure 7. It is seen that excessive accumulation of biofilms on the fluidized media leads to a greater bed height increase which, in turn, decreases the biofilm volume per unit fluidized bed volume (and therefore, the biomass concentration). Moreover, the impact of hydraulic loadings on the biomass concentration in an FBBR is more significant because higher hydraulic loadings lead to greater bed expansions. In extreme cases, wash-out of bioparticles from the reactor could occur.

C. Significance of Media Characteristics in an FBBR

The selection of media type in the design of a conventional biofilm system is relatively straightforward, because it is expected that the media selected would provide a high specific surface area for an extensive biofilm formation and an adequate void space to avoid bed clogging caused by biofilm accumulation. Other important characteristics include durability, strength, and cost. The plastic media developed in the early 1970s are commonly used today because they possess all the required characteristics.

The design engineer faces a somewhat different situation in the selection of media type for an FBBR. The fluidized state maintained in an FBBR allows the use of small media for a high specific surface area while avoiding high head loss and bed clogging problems which would be encountered in a conventional biofilm system such as a submerged filter. Nevertheless, such an advantage is partially offset by the fact that the hydrodynamic environment prevailing in an FBBR becomes a significant factor in the selection of media type. It is essential that the fluidization characteristics of an FBBR under anticipated operating conditions to be thoroughly evaluated before a given type of media is recommended. The following figures illustrate the significance of media characteristics on the performance of an FBBR.

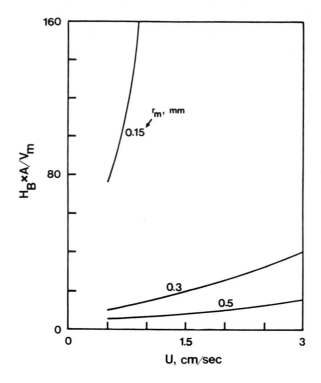

FIGURE 8.   Effect of media size on the FBBR bed expansion. ρ_m = 2.65 g/mℓ; δ = 100 μm; ρ = 65 mg/mℓ, and P = 0.93.

The effects of media characteristics on the FBBR bed expansion are illustrated in Figures 8 and 9. These figures show that the sensitivity of an FBBR toward variations in the hydraulic loading decreases with increased media size and density. This is extremely important with regard to the design and operation of an FBBR. In cases where small and/or light media are required to offset the higher energy requirements for fluidization of large and/or heavy media, then flow equilization provision prior to an FBBR should be considered in order to avoid undesirable washout of bioparticles.

Similar conclusions can be drawn on the effect of media size on the biomass concentration in an FBBR as illustrated in Figure 10. The importance of the prudent selection of media type is clearly demonstrated.

IV. DISCUSSION

A kinetic model based on the concept of effectiveness factor has been developed for a biofilm system as an alternative to the empirical models. This mechanistic model allows a clear definition of the kinetic regime of a biofilm system and thus a straightforward assessment of the effects of mass transfer limitations. A new concept which aids in understanding the performance characteristics of a biofilm system has also been defined in terms of effectiveness factor. It is the active biomass concentration within the reactor which is defined as the product of effectiveness factor and the total reactor biomass concentration. The sensitivity analysis performed utilizing the proposed kinetic model indicates that the biofilm thickness is the single most important parameter affecting the effectiveness factor and the biomass concentration and thus, the overall performance of a biofilm system.

A. Optimization of Conventional Biofilm Systems

The denitrification data illustrated in Figure 5 indicate that the biofilm thickness in a

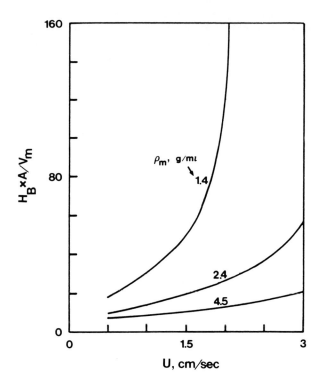

FIGURE 9. Effect of media density (ρ_m) on the FBBR bed expansion. $r_m = 0.2$ mm; $\delta = 100$ μm; $\rho = 65$ mg/mℓ, and P = 0.93.

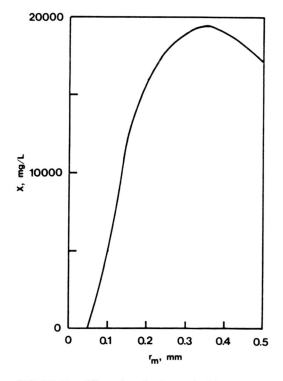

FIGURE 10. Effect of media size on the biomass concentration (X) in an FBBR. $\rho_m = 2.65$ g/mℓ, U = 1 cm/sec, $\rho = 65$ mg/mℓ, $\delta = 100$ μm, and P = 0.93.

conventional biofilm system should be maintained near or at the level for full substrate penetration (δ_ℓ) in order to insure optimal performance.[46] The δ_ℓ value for a specific application can be calculated directly from Equation 30. In practice, nevertheless, it may be difficult to maintain a close control in biofilm thickness because of the operational and structural characteristics involved. In a trickling filter, the biofilm thickness is naturally controlled by biofilm sloughing induced by the hydraulic shear force. It is surmised that, by maintaining the hydraulic loading at an adeqaute level through recirculation of filter effluent, the biofilm thickness may become more controllable. Further study in this aspect is recommended.

The biofilm thickness in an RBC is controlled by rotation of discs through the wastewater. Therefore, within practical limits, the biofilm thickness in an RBC is controllable by the engineer. In a submerged filter, periodic backwashing is commonly practiced to remove excessive biomass from the filter in order to avoid bed clogging. This practice can be used as a means to control the biofilm thickness within a desirable range. Moreoever, recirculation of filter effluent can also be used to maintain the hydraulic loading at an adequate level to continuously shear the biofilm from the media.

In addition to biofilm control, which is a process operation and control concern, maximization of active biomass concentration in a conventional biofilm system can also be realized in the design phase by using the media which provide a very high specific surface area. The plastic media available on the market are adequate in this aspect.

B. Optimization of FBBRs

Unlike the conventional biofilm system. optimization of an FBBR is not a straightforward process because many interrelated factors are involved. The biofilm thickness should be controlled at a level which yields the maximum active biomass concentration within the reactor. Maximization of either effectiveness factor or total biomass concentration does not necessarily yield the optimal performance of an FBBR. The desirable active biomass concentration for a specific application can be calculated via the kinetic model developed herein and the iterative procedure proposed by Mulcahy and LaMotta.[41]

Once a media type is selected in the design phase, FBBR optimziation is dependent on proper specification of media size and biofilm thickness. The relationship among media size (r_m), biofilm thickness (δ), and active biomass concentration (X_a) is shown in a three-dimensional diagram as illustrated in Figure 11. It is seen that the shape of the surface in Figure 11 is such that several r_m, δ values exist that maximize X_a. These r_m, δ values exist in the area where the surface curvature is small. Therefore, the design engineer can select a suitable r_m, δ value to meet his specific design and operation requirements while maintaining the performance of an FBBR at its optimal.

Control of the expanded FBBR bed height, via mechanical biofilm separation and subsequent biomass wasting, provides the most direct and convenient means to maintain the equilibrium biofilm thickness at a desirable level. In practice, the expanded bed height is not constantly controlled at a given level by continuously wasting overgrown bioparticles from the reactor. If it were, the separated biomass would be carried away as a dilute sludge requiring further thickening before final disposal. This inevitably would increase the costs associated with sludge disposal.

A more practical way of controlling the expanded bed height is to allow it to fluctuate over a range determined by the desirable expanded bed height and the expansion rate of the fluidized bed.[31] Shieh et al. reported that the biofilm thickness could be controlled at approximately 150 μm in an FBBR designed for carbon oxidation of a concentrated corn starch wastewater (influent $BOD_5 = 2000$ mg/ℓ) via this control strategy in spite of the fact that the substrate loadings were varied over a wide range.[31]

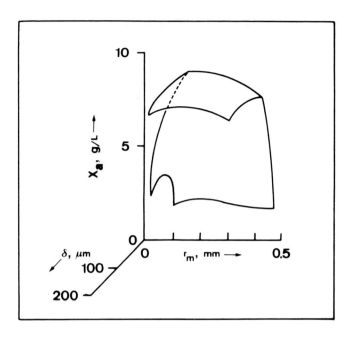

FIGURE 11. Combined effect of media size (r_m) and biofilm thickness (δ) on the active biomass concentration (X_a) in a denitrification FBBR. $k_o = 3.32 \times 10^{-5}\ell/\text{sec}$, $D_e = 0.825 \times 10^{-5}$ cm²/sec, $\rho = 65$ mg/mℓ; $\rho_m = 2.65$ g/mℓ, P = 0.93, and $S_b = 2$ mg NO$_3$-N/ℓ.

C. Evaluation of Intrinsic Kinetic Coefficients

In order to use the mechanistic biofilm model for process design, operation, and control, the intrinsic kinetic coefficients k_o and D_e and the microbial parameters ρ and P must be known. These coefficients and parameters can be evaluated experimentally using the experimental apparatus which allows a clear differentiation among the steps involved in the overall reaction. Mulcahy et al. have demonstrated that a rotating-disk biofilm reactor can be conveniently used for this purpose.[46] A brief description of the principle involved is presented below. Details of the experimental procedure are described elsewhere.[46]

If the biofilm thickness on the disk is equal or less than δ_ℓ, full substrate penetration of the biofilm will occur and thus the effectiveness factor is unity. Then, from Equations 9 and 15, it is clear that

$$\frac{(S_{bi} - S_b)}{aV} = \rho k_o \delta \tag{36}$$

Thus, the observed substrate conversion rate is a linear function of biofilm thickness with the slope (b) equal to ρk_o. Therefore:

$$k_o = \frac{b}{\rho} \tag{37}$$

If the biofilm thickness on the disk exceeds δ_ℓ, then the observed substrate conversion rate becomes independent of biofilm thickness and remains constant. Thus, the smallest biofilm thickness which yields a constant substrate conversion rate is δ_ℓ (or x_c) and from Equation 7 it is clear that:

$$D_e = \frac{\rho k_o \delta_\ell^2}{2S_b} \qquad (38)$$

where S_b is the bulk-liquid substrate concentration measured at the biofilm thickness δ_ℓ.

Thus, by running the rotating-disk biofilm reactor over a wide range of biofilm thicknesses, both k_o and D_e can be directly calculated. The kinetic coefficients measured via this experimental procedure are intrinsic ones which can be applied to predict the substrate conversion rates in other biofilm systems. Mulcahy et al. have shown that the NO_3-N concentration profiles through a plug-flow denitrification FBBR could be accurately predicted using k_o and D_e independently measured in a rotating-disk biofilm reactor operated under similar conditions.[46] Typical examples of their results are illustrated in Figure 12. The close agreement between observed and predicted values suggests that this experimental approach is viable for biofilm kinetic studies.

Tables 2 and 3 summarize values of k_o and D_e, respectively, for the intrinsic zero-order reactions that are of interest in wastewater treatment.[42,47]

D. Evaluation of Microbial Parameters

The microbial parameters ρ and P can be evaluated along with the intrinsic kinetic coefficients. A number of movable slides can be fitted on the disk so that they can be removed for biofilm thickness and biomass weight determinations. The biofilm thickness can be measured using a microscope equipped with a stage micrometer.[46] The dry weight of the biofilm scraped from the slide (W_b) can be determined as the weight of the total volatile solids. Then the biofilm dry density (ρ) can be calculated as:

$$\rho = \frac{W_b}{\delta A'} \qquad (39)$$

where A' is the surface area of the slide, L^2.

The biofilm moisture content (P) can be calculated as:

$$P = \frac{\rho_w - \rho}{\rho_w} \qquad (40)$$

where ρ_w is the biofilm wet density, M/L^3, which can be calculated as:

$$\rho_w = \frac{W'_b}{\delta A'} \qquad (41)$$

where W'_b is the wet weight of the biofilm on the slide, M.

Mulcahy et al. reported that the biofilm dry density values measured via this experimental approach were relatively constant in a denitrification biofilm system with biofilm thickness up to about 300 μm. Their results are illustrated in Figure 13.[46] Reported biofilm dry density values determined from different biofilm systems are summarized in Table 4.[31]

V. CONCLUDING REMARKS

An attempt has been made in this chapter to develop mechanistic models for biofilm systems designed for wastewater treatment. The concept of effectiveness factor has been introduced as a convenient means to define the kinetic regime of a biofilm system. The sensitivity analysis performed indicates that biofilm thickness is by far the most significant

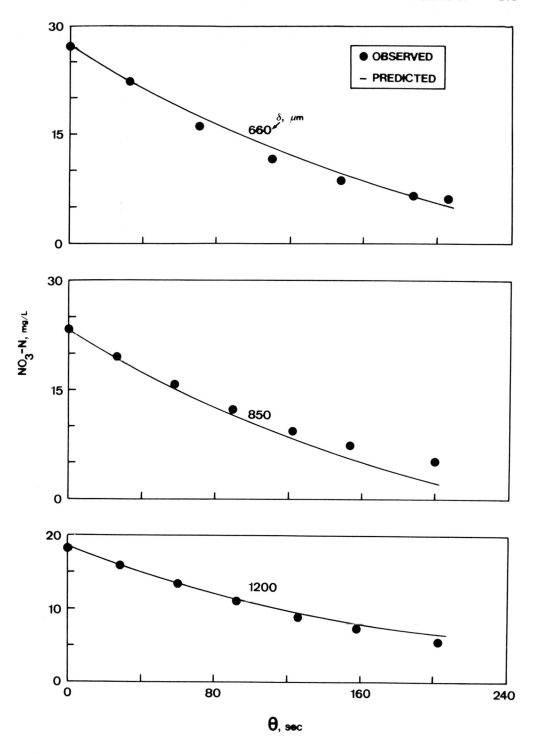

FIGURE 12. Comparison of predicted and observed NO_3-N profiles in a plug-flow denitrificaton FBBR.[46]

Table 2
INTRINSIC ZERO-ORDER REACTION RATE CONSTANTS[42,47]

Substrate	Electron acceptor	$k_o(10^{-5}\ell/sec)$	Remarks
NO_3-N	Methanol	3.32	20°C Nitrified secondary effluent
NH_4-N	O_2	8.67	20°C Synthetic wastewater
Glucose	O_2	7—50[a]	20°C Synthetic wastewater
Glucose	NO_3^-	3.6[a]	
NO_2-N	O_2	2.31	Pure culture

[a]   Units are in milligrams per hour per cubic centimeter.

Table 3
EFFECTIVE DIFFUSIVITIES IN BIOFILMS[42,47]

Substrate	Reactor configuration	Effective diffusivity ($10^{-5}cm^2/sec$)
Oxygen	Rotating tube	1.5
Oxygen	Plate	0.04
Oxygen	Trickling filter	0.82
Glucose	Annular reactor	0.28
Glucose	Inclinde glass plate	0.07
Ammonia	Inclined glass plate	1.7
Ammonia	Biofilm[a]	1.5
Nitrite	Biofilm[a]	1.4
Nitrate	Biofilm[a]	1.6
Nitrate	Rotating disk	0.82

[a]   Made by filtering dispersed microorganisms onto the support filter.

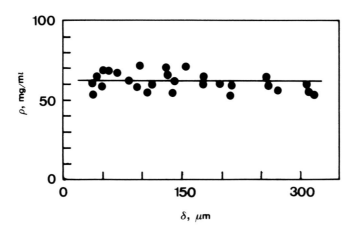

FIGURE 13.   Biofilm dry density (ρ) as a function of biofilm thickness (δ) in a denitrification biofilm system.[46]

Table 4
BIOFILM DRY DENSITY VALUES DETERMINED FROM DIFFERENT BIOFILM SYSTEMS[31]

Substrate	Reactor configuration	Biofilm dry density(g/mℓ)	Remarks
Glucose	Annular reactor	0.095	
Glucose	Tubular reactor	0.030	$S_{bi} = 20$ mg/ℓ
		0.040	$= 30$ mg/ℓ
		0.050	$= 70$ mg/ℓ
Glucose/nutrient broth/sodium acetate	Tubular reactor	0.105	$\delta = 100$ μm
		0.030	> 400 μm
Municipal wastewater	Tubular reactor	0.050	$\delta = 100$ μm
		0.045	$= 500$ μm
		0.030	$= 1100$ μm
Municipal wastewater	Inclined plate	0.033	
Municipal wastewater	Trickling filter	0.077	
Industrial wastewater[a]	RBC	0.028	$\delta = 1100$—3800 μm
Synthetic ammonium wastewater	Biofilm[b]	0.050—0.080	
Nitrified secondary effluent	FBBR	0.065	$\delta < 300$ μm
		0.030	> 630 μm
Corn starch wastewater	FBBR	0.072	$\delta = 150$ μm
		0.068	$= 173$ μm
Municipal wastewater	FBBR	0.075	$\delta = 89\mu$m
		0.048	$= 113$ μm
		0.039	$= 292$—312 μm

[a] Generated from an industrial process for recovery of urban solid wastes.
[b] Made by filtering dispersed microorganisms onto the support filter.

parameter affecting the overall performance of a biofilm system. Moreover, the media characteristics are also important, especially in the operation and control of an FBBR. The mechanistic models developed allow a clear differentiation among the steps involved in the overall reaction occurs in a biofilm system.

The application of the mechanistic model to the design, operation, and control of a biofilm system requires adequate knowledge of the intrinsic kinetics and the hydrodynamics of the biofilm system involved. Unfortunately, there is a paucity of information in these aspects. As a result, more research efforts are needed to gain better insight into the complex nature of the biofilm system. Further research in the following areas is recommended:

1. Development of mechanistic biofilm models and evaluation of intrinsic kinetic coefficients and microbial parameters for other types of reactions commonly encountered in wastewater treatment.
2. Evaluation of the effects of environmental conditions such as pH and temperature on the intrinsic kinetic coefficients and microbial parameters.
3. Development of mechanistic biofilm models which describe reaction under multiple or sequential substrate control.
4. Incorporation of mechanistic biofilm models in the overall process control strategies of the biofilm systems.

REFERENCES

1. **Metcalf and Eddy, Inc.,** *Wastewater Engineering: Treatment, Disposal, Reuse,* 2nd ed., McGraw-Hill, New York, 1979, chap. 9.
2. **LaMotta, E. J.,** Evaluation of Diffusional Resistances in Substrate Utilization by Biological Films, Ph.D. dissertation, University of North Carolina, Chapel Hill, N.C., 1974.
3. **Smith, E. D. and Bandy, J. T.,** State of knowledge for rotating biological contactor technology, in *Proc. 1st Int. Conf. Fixed-Film Biological Processes,* Wu, Y.C., Smith, E. D., Miller, R. D., and Patken, E. J. O., Eds., University of Pittsburgh, Pittsburgh, Pa., 1982, 1.
4. **Sachs, E. F., Jennett, J. C., and Rand, M. C.,** Pharmaceutical waste treatment by anaerobic filter, *J. Environ. Eng. Div.,* 108, EE2, 297, 1982.
5. **Jennett, J. C. and Dennis, N. D., Jr.,** Anaerobic filter treatment of pharmaceutical waste, *J. Water Pollut. Control Fed.,* 47, 104, 1975.
6. **Mueller, J. A. and Mancini, J. L.,** Anaerobic filter kinetics and applications, in *Proc. 30th Purdue Ind. Waste Conf.,* Purdue University, West Lafayette, Ind., 1975, 423.
7. **Young, J. C. and McCarty, P. L.,** The anaerobic filter for wastewater treatment in *Proc. 22nd Purdue Ind. Waste Conf.,* Purdue University, West Lafayette, Ind., 1967, 559.
8. **Khan, K. A., Suidan, M. T., and Cross, W. H.,** Anaerobic activated carbon filter for the treatment of phenol-bearing wastewater, *J. Water Pollut. Control Fed.,* 53, 1519, 1981.
9. **Obayashi, A. W., Stensel, H. D., and Kominek, E.,** Anaerobic treatment of high-strength wastes, *Chem. Eng. Prog.,* 68, 1981.
10. **Suidan, M. T., Cross, W. H., and Fong, M.,** Continuous bioregeneration of granular activated carbon during the anaerobic degradation of catechol, *Prog. Water Technol.,* 12, 203, 1980.
11. **Arora, H. C., Chattopadhya, S. N., and Routh, T.,** Treatment of vegetable tanning effluent by the anaerobic contact filter process, *Water Pollut. Control Fed.,* 584, 1975.
12. **Harremoës, P.,** The significance of pore diffusion to filter denitrification, *J. Water Pollut. Control Fed.,* 48, 377, 1976.
13. **Harremoës, P. and Reimer, M.,** Pilot-scale experiments on down-flow filter denitrification, *Prog. Water Technol.,* 8, 4/5, 557, 1977.
14. **Murphy, K. L. and Sutton, P. M.,** Pilot scale studies on biological denitrification, *Prog. Water Technol.,* 7, 2, 317, 1975.
15. **Iida, Y. and Teranishi, A.,** Nitrogen removal from municipal wastewater by a single submerged filter, *J. Water Pollut. Control Fed.,* 56, 251, 1984.
16. **Shieh, W. K. and Keenan, J. D.,** Fluidized bed biofilm reactor for wastewater treatment, in *Advances in Biochemical Engineering,* Fiechter, A., Ed., Springer-Verlag, Berlin, 1986.
17. **Hoyland, G.,** Mass-transfer model for aeration, *Prog. Water Technol.,* 11, 3, 237, 1979.
18. **Roberts, J.,** Towards a better understanding of high rate biological film flow reactor theory, *Water Res.,* 7, 1561, 1973.
19. **Grady, C. P. L., Jr.,** Modeling of biological fixed-films — a state-of-the-art review, in *Proc. 1st Int. Conf. Fixed-Film Biological Processes,* Wu, Y. C., Smith, E. D., Miller, R. D., and Patken, E. J. O., Eds., University of Pittsburgh, Pittsburgh, Pa., 1982, 344.
20. **Shieh, W. K., Mulcahy, L. T., and LaMotta, E. J.,** Mathematical model for the fluidized bed biofilm reactor, *Enzyme Microb. Technol.,* 4, 269, 1982.
21. **Shieh, W. K.,** Mass transfer in a rotating biological contactor, *Water Res.,* 16, 1071, 1982.
22. **Shieh, W. K.,** Suggested kinetic model for the fluidized bed biofilm reactor, *Biotechnol. Bioeng.,* 24, 667, 1980.
23. **Rittman, B. E.,** Comparative performance of biofilm reactor types, *Biotechnol. Bioeng.,* 24, 1341, 1982.
24. **Mulcahy, L. T., Shieh, W. K., and LaMotta, E. J.,** Simplified mathematical model for a fluidized bed biofilm reactor, *Water — 1980,* AIChE Symp. Ser., 77, 209, 273, 1981.
25. **LaMotta, E. J.,** Internal diffusion and reaction in biological films, *Environ. Sci. Technol.,* 10, 8, 765, 1976.
26. **Shieh, W. K. and Mulcahy, L. T.,** FBBR kinetics — a rational design and optimization approach, in *Anaerobic Treatment of Wastewater in Fixed Film Reactors,* Henze, M., Ed., Pergamon Press, Oxford, 1983, 321.
27. **Harremoës, P.,** Criteria for nitrification in fixed film reactors, *Water Sci. Technol.,* 14, 167, 1872.
28. **Atkinson, B. and Mavituna, F.,** *Biochemical Engineering and Biotechnology Handbook* Macmillan, Surrey, 1983, chaps. 7 and 10.
29. **Shieh, W. K., Mulcahy, L. T., and LaMotta, E. J.,** Fluidized bed biofilm reactor effectiveness factor expressions, *Trans. Inst. Chem. Eng.,* 59, 129, 1981.
30. **Sutton, P. M., Shieh, W. K., Kos, P., and Dunning, P. R.,** Dorr-Oliver's Oxitron System™ fluidized-bed water and wastewater treatment process, in *Biological Fluidized Bed Treatment of Water and Wastewater,* Cooper, P. F. and Atkinson, B., Eds., Ellis Horwood, Chichester, 1981, chap. 17.

31. **Shieh, W. K., Sutton, P. M., and Kos, P.,** Predicting reactor biomass concentration in a fluidized-bed system, *J. Water Pollut. Control Fed.,* 53, 1574, 198.
32. **Nutt, S. G., Stephenson, J. P., and Pries, J. H.,** Nitrification kinetics in the biological fluidized bed process, presented at the 53rd Annual Conference of the Water Pollution Control Federation, Las Vegas, 1980.
33. **Rusten, B.,** Wastewater treatment with aerated submerged biological filters, *J. Water Pollut. Control. Fed.,* 56, 424, 1984.
34. **Shieh, W. K. and Chen, C. Y.,** Biomass hold-up correlations for a fluidized bed biofilm reactor, *Chem. Eng. Res. Des.,* 62, 133, 1984.
35. **Webb, C., Black, G. M., and Atkinson, B.,** Liquid fluidization characteristics of biological beds, *Biotechnol. Bioeng.,* 25, 1321, 1983.
37. **Richardson, J. F. and Zaki, W. N.,** Sedimentation and fluidization. I, *Trans. Inst. Chem. Eng.,* 32, 35, 1954.
38. **Lewis, E. W. and Bowerman, E. W.,** Fluidization of solid particles in liquid, *Chem. Eng. Prog.,* 48, 605, 1952.
39. **Wen, C. Y. and Yu, Y. H.,** Mechanics of fluidization, *Chem. Eng. Prog. Symp. Ser.,* 62, 100, 1962.
40. **Chen, P. and Pei, D. C. T.,** Fluidization characteristics of fine particles, *Can. J. Chem. Eng.,* 62, 464, 1984.
41. **Mulcahy, L. T. and LaMotta, E. J.,** Mathematical model of the fluidized bed biofilm reactor, Report No. Env. E. 59-78-2, Department of Civil Engineering, University of Massachusetts, Amherst, Mass., 1978.
42. **Harremoës, P.,** Biofilm kinetics, in *Water Pollution Microbiology,* Vol. 2, Mitchell, R., Ed., Wiley-Interscience, New York, 1978, chap. 4.
43. **O'Shaughnessy, J. C. and Blanc, F. C.,** Biological nitrification and denitrification using rotating biological contactors, Publ. No. 97, Water Resource Research Center, University of Massachusetts, Amherst, Mass., 1978.
44. **Soyupak, S., Murphy, K. L., and Jank, B. E.,** Analysis of denitrification with a complete submerged pilot-scale rotating biological contactor, *Biotechnol. Bioeng.,* 21, 1787, 1979.
45. **Wanatabe, Y. and Ishiguro, M.,** Denitrification kinetics in a submerged rotating biological disk unit, *Prog. Water Technol.,* 10, 5/6, 187, 1978.
46. **Mulcahy, L. T., Shieh, W. K., and LaMotta, E. J.,** Experimental determination of intrinsic denitrification kinetic constants, *Biotechnol. Bioeng.,* 23, 2403, 1981.
47. **Shieh, W. K.,** Kinetics of Simultaneous Diffusion and Reaction for the Nitrification Process in Suspended Growth Systems, Ph.D. dissertation, University of Massachusetts, Amherst, Mass., 1978.

INDEX

A

M

N

O

P